501
GEOMETRY
QUESTIONS

501 GEOMETRY QUESTIONS

2nd Edition

LEARNINGEXPRESS®

NEW YORK

ISBN 978-1-57685-894-3

Printed in the United States of America

9 8 7 6 5 4 3 2 1

Second Edition

For more information or to place an order, contact LearningExpress at:
    2 Rector Street
    26th Floor
    New York, NY 10006

Or visit us at:
    www.learningexpressllc.com

Contents

Introduction

Geometry is the study of figures in space. As you study geometry, you will measure these figures and determine how they relate to each other and the space they are in. To work with geometry you must understand the difference between representations on the page and the figures they symbolize. What you see is not always what is there. In space, lines define a square; on the page, four distinct black marks define a square. What is the difference? On the page, lines are visible. In space, lines are invisible because lines do not occupy space, in and of themselves. Let this be your first lesson in geometry: appearances may deceive.

Sadly, for those of you who love the challenge of proving the validity of geometric postulates and theorems—the statements that define the rules of geometry—this book is not for you. It will not address geometric proofs or zigzag through tricky logic problems, but it will focus on the practical application of geometry towards solving planar (two-dimensional) spatial puzzles. As you use this book, you will work under the assumption that every definition, every postulate, and every theorem is "infallibly" true.

How to Use This Book

Review the introduction to each chapter before answering the questions in that chapter. Problems toward the end of this book will demand that you apply multiple lessons to solve a question, so be sure you know the preceding chapters well. Take your time; refer to the introductions of each chapter as frequently as you need to, and be sure you understand the answer explanations at the end of each section. This book provides the practice; you provide the initiative and perseverance.

Author's Note

Some geometry books read like instructions on how to launch satellites into space. While geometry is essential to launching NASA space probes, a geometry book should read like instructions on how to make a peanut butter and jelly sandwich. It's not that hard, and after you are done, you should be able to enjoy the product of your labor. Work through this book, enjoy some pb & j, and soon you too can launch space missions if you want.

501 GEOMETRY QUESTIONS

1

The Basic Building Blocks of Geometry

Before you can tackle geometry's toughest concepts, you must first understand the building blocks of geometry: the point, the line, and the plane. Points, lines, and planes are critical to geometry because they combine to create the angles, polygons, prisms, and circles that are the focus of geometry. Let's begin by learning the definitions of the three rudimentary elements of geometry.

Point

Point A

• A   • A

Figure   Symbol

A **point** is a location in space; it indicates position. It occupies no space of its own, and it has no dimension of its own. One of the most important concepts to remember about points is that it only takes two points to define, or create, a line.

Line

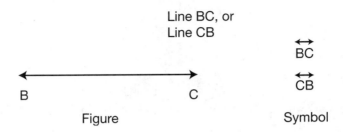

Line BC, or
Line CB

$\overset{\leftrightarrow}{BC}$

$\overset{\leftrightarrow}{CB}$

B   C

Figure   Symbol

A **line** is a set of continuous points infinitely extending in opposite directions. It has infinite length, but no depth or width. A line is most commonly defined by its endpoints, but any two points on a line can be used to name it.

Plane

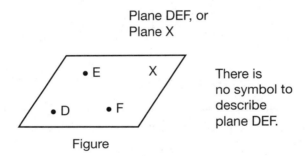

Plane DEF, or
Plane X

• E   X

• D   • F

There is
no symbol to
describe
plane DEF.

Figure

A **plane** is a flat expanse of points expanding in every direction. Planes have two dimensions: length and width. They do not have depth.

As you probably noticed, each "definition" above builds upon the "definition" before it. There is the point; then there is a series of points; then there is an expanse of points. In geometry, space is pixilated much like the image you see on a TV screen. Be aware that definitions from this point on will build upon each other much like these first three definitions.

Collinear/Noncollinear

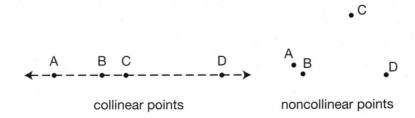

collinear points   noncollinear points

Collinear points are points that form a single straight line when they are connected (two points are always collinear). **Noncollinear points** are points that do not form a single straight line when they are connected (only three or more points can be noncollinear).

Coplanar/Noncoplanar

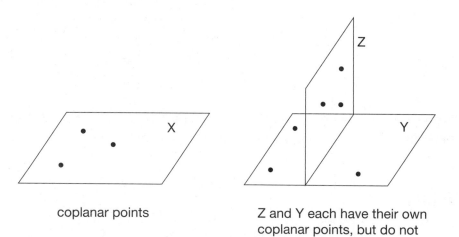

coplanar points

Z and Y each have their own coplanar points, but do not share coplanar points.

Coplanar points are points that occupy the same plane. **Noncoplanar points** are points that do not occupy the same plane.

Ray

Ray GH

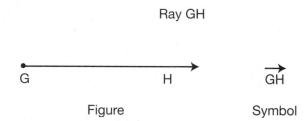

Figure   Symbol

A **ray** begins at a point (called an *endpoint* because it marks the *end* of a ray), and infinitely extends in one direction.

Opposite Rays

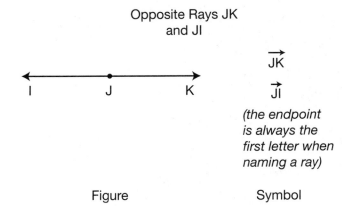

Opposite Rays JK
and JI

$\overrightarrow{JK}$

$\overrightarrow{JI}$

(the endpoint is always the first letter when naming a ray)

Figure   Symbol

Opposite rays are rays that share an endpoint and infinitely extend in opposite directions. Opposite rays form straight angles.

Angles

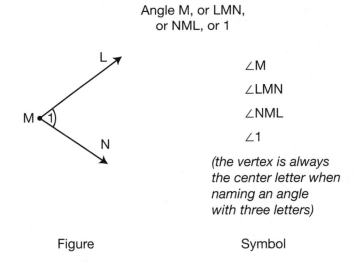

Angle M, or LMN,
or NML, or 1

∠M

∠LMN

∠NML

∠1

(the vertex is always the center letter when naming an angle with three letters)

Figure   Symbol

Angles are made up of two rays that share an endpoint but infinitely extend in different directions.

Line Segment

Line Segment OP,
or PO

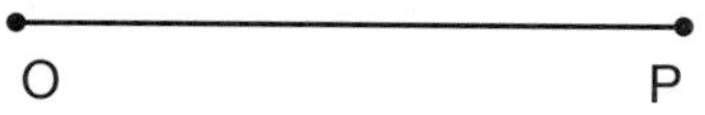

| O | P |

Figure   Symbol

$\overline{OP}$

$\overline{PO}$

A **line segment** is part of a line with two endpoints. Although not infinitely extending in either direction, the line segment has an infinite set of points between its endpoints.

Set 1

Answer the following questions.

1. What are the three building blocks of geometry? _____, _____, _____

2. Two _____ define a line.

3. What does a point indicate? _____

4. How far does a ray extend? _____

5. How many non-collinear points does it take to determine a plane? _____

6. An angle is made up of two _____ that share a common _____.

7. How does a line segment differ from a line? _____ _____

Set 2

Choose the best answer.

8. Collinear points
   a. determine a plane.
   b. are circular.
   c. are noncoplanar.
   d. are coplanar.

9. How many distinct lines can be drawn through two points?
   a. 0
   b. 1
   c. 2
   d. an infinite number of lines

10. Lines are always
    a. solid.
    b. finite.
    c. noncollinear.
    d. straight.

11. The shortest distance between any two points is
    a. a plane.
    b. a line segment.
    c. a ray.
    d. an arch.

12. Which choice below has the most points?
    a. a line
    b. a line segment
    c. a ray
    d. No determination can be made.

Set 3

Answer questions 13 through 16 using the figure below.

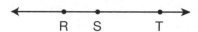

13. Write three different ways to name the line above. Are there still other ways to name the line? If there are, what are they? If there aren't, why not?

14. Name four different rays. Are there other ways to name each ray? If there are, what are they? If there aren't, why not?

15. Name a pair of opposite rays. Are there other pairs of opposite rays? If there are, what are they?

16. Name three different line segments. Are there other ways to name each line segment? If there are, what are they? If there aren't, why not?

Set 4

Answer questions 17 through 20 using the figure below.

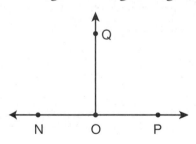

17. Write three different ways to name the horizontal line above. Are there still other ways to name the line? If there are, what are they? If there aren't, why not?

18. Name five different rays contained in the figure above. Are there other ways to name each ray? If there are, what are they? If there aren't, why not?

19. Name a pair of opposite rays. Are there other pairs of opposite rays? If there are, what are they?

20. Name three angles. Are there other ways to name each angle? If there are, what are they? If there aren't, why not?

Set 5

Answer questions 21 through 23 using the figure below.

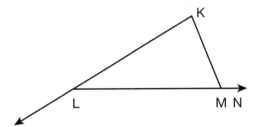

21. Name three different rays. Are there other rays? If there are, what are they?

22. Name five angles. Are there other ways to name each angle? If there are, what are they? If there aren't, why not?

23. Name five different line segments. Are there other ways to name each line segment? If there are, what are they? If there aren't, why not?

Set 6

Ann, Bill, Carl, and Dan work in the same office building. Dan works in the basement while Ann, Bill, and Carl share an office on level X. At any given moment of the day, they are all typing at their desks. Bill likes a window seat; Ann likes to be near the bathroom; and Carl prefers a seat next to the door. Their three cubicles do not line up.

Answer the following questions using the description above.

24. Level X can also be called
    a. Plane Ann, Bill, and Carl.
    b. Plane Ann and Bill.
    c. Plane Dan.
    d. Plane Carl, X, and Bill.

25. If level X represents a plane, then level X has
    a. no points.
    b. only three points.
    c. a finite set of points.
    d. an infinite set of points extending infinitely.

26. If Ann and Bill represent points, then Point Ann
    a. has depth and length, but no width; and is noncollinear with point Bill.
    b. has depth, but no length and width; and is noncollinear with point Bill.
    c. has depth, but no length and width; and is collinear with point Bill.
    d. has no depth, length, and width; and is collinear with point Bill.

27. If Ann, Bill, and Carl represent points, then Points Ann, Bill, and Carl are
    a. collinear and noncoplanar.
    b. noncollinear and coplanar.
    c. noncollinear and noncoplanar.
    d. collinear and coplanar.

28. A line segment drawn between Carl and Dan is
   a. collinear and noncoplanar.
   b. noncollinear and coplanar.
   c. noncollinear and noncoplanar.
   d. collinear and coplanar.

Answers

Set 1

1. The three building blocks of geometry are **points**, **lines**, and **planes**, which are used to create the more complex shapes and figures studied in geometry.

2. A line contains an infinite number of points, but it only takes two **points** to define a line.

3. A point indicates **position**.

4. A ray starts at a point and extends **infinitely**. It contains an infinite number of points.

5. While two points determine a line, it takes **three** non-collinear points determine a plane.

6. An angle is made up of two **rays** that share a common **endpoint**.

7. A line segment is **part** of a line that has two endpoints. A line extends infinitely in both directions.

Set 2

8. **d.** Collinear points are also coplanar. Choice **a** is not the answer because noncollinear points determine planes, not a single line of collinear points.

9. **b.** An infinite number of lines can be drawn through one point, but only one straight line can be drawn through two points.

10. **d.** Always assume that in plane geometry a line is a straight line unless otherwise stated. Process of elimination works well with this question: Lines have one dimension, length, and no substance; they are definitely not solid. Lines extend to infinity; they are not finite. Finally, we defined noncollinear as a set of points that "do

not line up"; we take our cue from the last part of that statement. Choice **c** is not our answer.

11. **b.** A *line segment* is the shortest distance between any two points.

12. **d.** A line, a line segment, and a ray are sets of points. How many points make a set? An infinite number. Since a limit cannot be put on infinity, not one of the answer choices has more than the other.

Set 3

13. Any six of these names correctly describe the line: $\overleftrightarrow{RS}$, $\overleftrightarrow{SR}$, $\overleftrightarrow{RT}$, $\overleftrightarrow{TR}$, $\overleftrightarrow{ST}$, $\overleftrightarrow{TS}$, $\overleftrightarrow{RST}$, and $\overleftrightarrow{TSR}$. Any two points on a given line, regardless of their order, describes that line. Three points can describe a line, as well.

14. Two of the four rays can each be called by only one name: $\overrightarrow{ST}$ and $\overrightarrow{SR}$. Ray names $\overrightarrow{RT}$ and $\overrightarrow{RS}$ are interchangeable, as are ray names $\overrightarrow{TS}$ and $\overrightarrow{TR}$; each pair describes one ray. $\overrightarrow{RT}$ and $\overrightarrow{RS}$ describe a ray beginning at endpoint R and extending infinitely through •T and •S. $\overrightarrow{TS}$ and $\overrightarrow{TR}$ describe a ray beginning at endpoint T and extending infinitely through •S and •R.

15. $\overrightarrow{SR}$ and $\overrightarrow{ST}$ are opposite rays. Of the four rays listed, they are the only pair of opposite rays; they share an endpoint and extend infinitely in opposite directions.

16. Line segments have two endpoints and can go by two names. It does not matter which endpoint comes first. $\overline{RT}$ is $\overline{TR}$; $\overline{RS}$ is $\overline{SR}$; and $\overline{ST}$ is $\overline{TS}$.

Set 4

17. Any six of these names correctly describes the horizontal line: $\overleftrightarrow{NP}$, $\overleftrightarrow{PN}$, $\overleftrightarrow{NO}$, $\overleftrightarrow{ON}$, $\overleftrightarrow{PO}$, $\overleftrightarrow{OP}$, $\overleftrightarrow{NOP}$, $\overleftrightarrow{PON}$. Any two points on a given line, regardless of their order, describe that line.

18. Three of the five rays can each be called by only one name: $\overrightarrow{OP}$, $\overrightarrow{ON}$, and $\overrightarrow{OQ}$. Ray-names $\overrightarrow{NO}$ and $\overrightarrow{NP}$ are interchangeable, as are ray names $\overrightarrow{PO}$ and $\overrightarrow{PN}$; each pair describes one ray each. $\overrightarrow{NO}$ and $\overrightarrow{NP}$ describe a ray beginning at endpoint N and extending infinitely through •O and •P. $\overrightarrow{PO}$ and $\overrightarrow{PN}$ describe a ray beginning at endpoint P and extending infinitely through •O and •N.

19. $\overrightarrow{ON}$ and $\overrightarrow{OP}$ are opposite rays. Of the five rays listed, they are the only pair of opposite rays; they share an endpoint and extend infinitely in opposite directions.

20. Angles have two sides, and unless a number is given to describe the angle, angles can have two names. In our case ∠NOQ is the same as ∠QON; ∠POQ is the same as ∠QOP; and ∠NOP is the same as ∠PON (in case you missed this one, ∠NOP is a straight angle). Letter O cannot by itself name any of these angles because all three angles share •O as their vertex.

Set 5

21. Two of the three rays can each be called by only one name: $\overrightarrow{KL}$ and $\overrightarrow{MN}$. $\overrightarrow{LN}$ and $\overrightarrow{LM}$ are interchangeable because they both describe a ray beginning at endpoint L and extending infinitely through •M and •N.

22. Two of the five angles can go by three different names. ∠KLM is the same as ∠NLK and ∠MLK. ∠LKM is ∠MKL and also ∠K. The other three angles can only go by two names each. ∠KMN is ∠NMK. ∠KML is ∠LMK. ∠LMN is ∠NML. Letter M cannot by itself name any of these angles because all three angles share •M as their vertex.

23. Line segments have two endpoints and can go by two names. It makes no difference which endpoint comes first. $\overline{LM}$ is $\overline{ML}$; $\overline{MN}$ is $\overline{NM}$; $\overline{LN}$ is $\overline{NL}$; $\overline{KM}$ is $\overline{MK}$; $\overline{KL}$ is $\overline{LK}$.

Set 6

24. **a.** Three noncollinear points determine a plane. In this case, we know level X is a plane and Ann, Bill, and Carl represent points on that plane. Ann and Bill together are not enough points to define the plane; Dan isn't on plane X and choice **d** doesn't make sense. Choice **a** is the only option.

25. **d.** Unlike a plane, an office floor can hold only so many people; however, imagine the office floor extending infinitely in every direction. How many people could it hold? An infinite number.

26. **d.** Just as the office floor can represent a plane, Ann and Bill can represent points. They acquire the characteristics of a point; and as we know, points have no dimension, and two points make a line.

27. **b.** Ann, Bill, and Carl are all on the same floor, which means they are all on the same plane, and they are not lined up. That makes them noncollinear but coplanar.

28. **d.** Carl and Dan represent two points; two points make a line; and all lines are collinear and coplanar. Granted, Dan and Carl are on two different floors; but remember points exist simultaneously on multiple planes.

2

Types of Angles

"If there is any one secret of success,
it lies in the ability to get the other person's point of view
and see things from that person's angle as well as from your own."
—Henry Ford

When Henry Ford talks about understanding some else's "angle," he is referring to their perspective. There are an infinite number of perspectives that exist, and the number of angles that can exist is also infinite.

What is *not* infinite, however, is how we categorize angles. In this chapter, you will learn how to classify all types of angles by their measurements.

The Parts of an Angle

Remember chapter 1 defined an **angle** as two rays sharing an endpoint and extending infinitely in different directions. The common endpoint is called the **vertex**. The rays are each called **sides**.

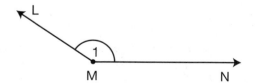

• M is a vertex
$\overrightarrow{ML}$ is a side
$\overrightarrow{MN}$ is another side

Classifying Angles

Angles are measured in degrees, which are a measurement of how much one side of an angle is rotated away from the other side. Degrees do not measure distance or length. The symbol used for degrees is °. Remember that angles are usually named with three letters, starting with a point on one side, followed by the vertex point, and ending with a point on the other side. It doesn't matter which side is named first. The following angle can be named ∠RSB or ∠BSR.

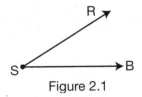

Figure 2.1

If a given vertex is a vertex for only one angle, than that point can be used to name the angle as well. For example, in the previous figure, ∠RSB can also be called ∠S. In the following angle, ∠A cannot be used as a name because point A is the vertex for ∠KAS as well as for ∠KAM. Therefore if ∠A was used, it would not be clear which angle was being referred to.

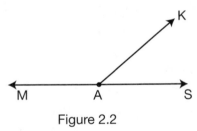

Figure 2.2

Angles are classified by their relationship to the degree measures of 90° and 180°:

Straight angles: An angle that measures 180° will look like a straight line and is called a **straight angle.**

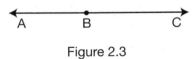

Figure 2.3

Right angles: An angle that measures 90° is called a **right angle.** Right angles have their own symbol to show they are 90°. A small square is drawn at the vertex of right angles to note their unique nature, as shown in Figure 2.4.

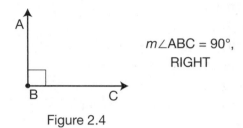

$m\angle ABC = 90°$,
RIGHT

Figure 2.4

Acute angles: An angle that is larger than 0° but smaller than 90° is called an **acute angle**. (A clever way to remember this is to think, "What a **cute** little angle!")

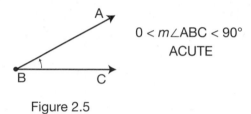

$0 < m\angle ABC < 90°$
ACUTE

Figure 2.5

Obtuse angles: An angle that is larger than 90° but smaller than 180° is called an **obtuse angle**. (Do you like rhymes? "That **obtuse** is as big as a **moose**!")

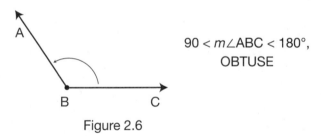

$90 < m\angle ABC < 180°$,
OBTUSE

Figure 2.6

Reflex angle: An angle that is larger than 180° but smaller than 360° is called a **reflexive angle**. There are no names for angles larger than 360°.

$180 < m\angle ABC < 360°$,
REFLEX

Figure 2.7

Set 7

Choose the best answer.

29. Angles that share a common vertex point **cannot**
    a. share a common angle side.
    b. be right angles.
    c. use the vertex letter name as an angle name.
    d. share interior points.

Choose the answer that *incorrectly* names an angle in each preceding figure.

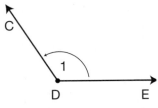

30. a. ∠CDE
    b. ∠CED
    c. ∠D
    d. ∠1

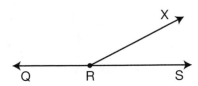

31. a. ∠R
    b. ∠QRS
    c. ∠XRS
    d. ∠XRQ

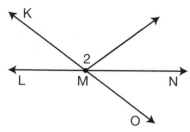

32. a. ∠KMN
   b. ∠NMO
   c. ∠KML
   d. ∠M

Set 8

Choose the best answer.

33. **True** or **False:** An acute angle plus an acute angle will always give you an obtuse angle.

34. "A straight angle minus an obtuse angle will give you an acute angle." Is this statement **sometimes**, **always**, or **never** true?

35. "A right angle plus an acute angle will give you an obtuse angle." Is this statement **sometimes**, **always**, or **never** true?

36. "An obtuse angle minus an acute angle will give you an acute angle." Is this statement **sometimes**, **always**, or **never** true?

Set 9

Label each angle measurement as acute, right, obtuse, straight, or reflexive.

37. 13.5°

38. 90°

39. 246°

40. 180°

41.

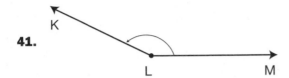

42.

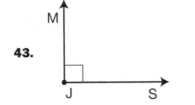

43.

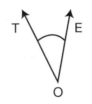

Set 10

For each diagram in this set, name every angle in as many ways as you can. Then label each angle as acute, right, obtuse, straight, or reflexive.

44.

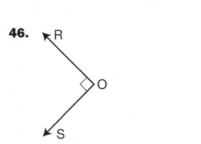

45.

46.

47.

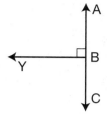

48.

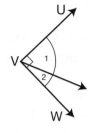

49.

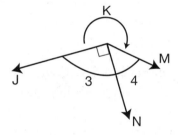

50.

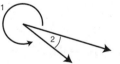

Answers

Set 7

29. **c.** If a vertex is shared by more than one angle, then it cannot be used to name any of the angles.

30. **b.** ∠CED describes an angle whose vertex is •E, not •D.

31. **a.** If a vertex is shared by more than one angle, then the letter describing the vertex cannot be used to name any of the angles. It would be too confusing.

32. **d.** If a vertex is shared by more than one angle, then the letter describing the vertex cannot be used to name any of the angles. It would be too confusing.

Set 8

33. **False.** Two small acute angles do not have to have a sum greater than 90°. For example, 10° + 20° = 30°, which is still acute.

34. **Always.** Since a straight angle equals 180°, and an obtuse angle is between 90° and 180°, the difference of a straight angle and an obtuse angle will always be between 0° and 90°.

35. **Always.** A right angle is 90° and an acute angle is greater than 0°and less than 90°. Therefore, the sum of a right angle and an acute angle will always be greater than 90° but less than 180°.

36. **Sometimes.** An obtuse angle that measures 95° minus an acute angle that measures 20° will result in an angle that measures 75° and is acute. However, the same obtuse angle that measures 95° minus an acute angle that measures 4° will result in an angle that measures 91° and is obtuse. Lastly, the same obtuse angle that measures 95° minus an acute angle that measures 5° will result in an angle that measures 90° and is a right angle.

Set 9

37. **Acute.** $0° < 13.5° < 90°$.

38. **Right angle.** $90°$.

39. **Reflexive.** $180° < 246° < 360°$.

40. **Straight angle.** $180°$.

41. **Obtuse.** $\angle KLM$ is greater than $90°$ and less than $180°$.

42. **Straight.** $\angle KLM$ is a straight line which measures $180°$.

43. **Right.** $\angle MJS$ is equal $90°$.

Set 10

44. **Acute.** $\angle TOE$, $\angle EOT$, or $\angle O$.

45. **Obtuse.** $\angle 1$.

46. **Right.** $\angle ROS$, $\angle SOR$, or $\angle O$.

47. **Right.** $\angle ABY$ or $\angle YBA$.
   Right. $\angle YBC$ or $\angle CBY$.
   Straight. $\angle ABC$ and $\angle CBA$.

48. **Acute.** $\angle 1$.
   Acute. $\angle 2$.
   Right. $\angle UVW$ or $\angle WVU$.

49. Since the vertex point is not labeled, the only angle that can be named by letter using a letter is $\angle K$, which is reflexive because it is greater than $180°$ and less than $360°$
   Right. $\angle 3$.
   Acute. $\angle 4$.

50. **Reflexive.** $\angle 1$.
   Acute. $\angle 2$.

3

Working with Lines

Every angle or shape that we come across in geometry or in the real world is made up of intersecting lines. However, there are also two special cases where lines do **not** intersect.

Parallel lines are coplanar lines that never intersect. Instead, they travel similar paths at the same distance from each other at all times. Lines **a** and **b** below in Figure 3.1 are parallel; the symbol used to represent parallel lines is **a ‖ b**. Another symbol used to indicate when two lines are parallel are the corresponding arrows you see in the figure. Sometimes corresponding tick marks are used instead of arrows.

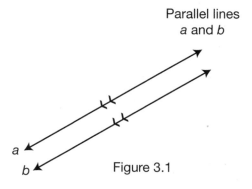

Parallel lines
a and *b*

a
b   Figure 3.1

Skew lines are noncoplanar lines that never intersect. They exist in two different planes, and travel dissimilar paths that will never cross. This is hard to draw, because it is hard to represent different planes in a two-dimensional drawing, but lines **a** and **b** below are skew. There is no symbol to represent skew lines.

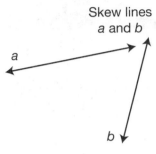

Figure 3.2

Intersecting Lines

When two lines are not parallel or skew, they will intersect. When two lines intersect, the point of the intersection is a point that exists on both lines—it is the only point that the lines have in common. Two intersecting lines create four angles. In Figure 3.3 below, line **a** and line **b** intersect at point **c**, creating the four different angles, **d**, **e**, **f**, and **g**. When two lines intersect, the sum of the four angles they create is 360°. You can tell this is true by looking at line **b**: angles **e** and **f** make a straight angle *above* line **b**, so they add to 180°, and angles **d** and **g** make a straight angle *below* line **b**, so they also add to 180°. Therefore, the four angles together add to 360°.

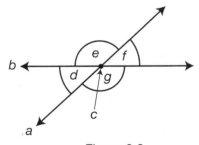

Figure 3.3

Now that you understand how intersecting lines create angles, it is time to learn about the special relationships that can exist between angles. When the sum of the measurement of any two angles equals 180°, the angles are called **supplementary angles.**

When straight lines intersect, the two angles next to each other are called **adjacent angles.** They share a vertex, a side, and no interior points. Adjacent angles along a straight line are always supplementary.

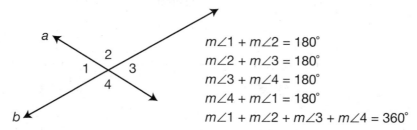

$$m\angle1 + m\angle2 = 180°$$
$$m\angle2 + m\angle3 = 180°$$
$$m\angle3 + m\angle4 = 180°$$
$$m\angle4 + m\angle1 = 180°$$
$$m\angle1 + m\angle2 + m\angle3 + m\angle4 = 360°$$

When straight lines intersect, opposite angles, or angles nonadjacent to each other, are called **vertical angles**. They are always congruent or equal in measure. The symbol for congruent is ≅.

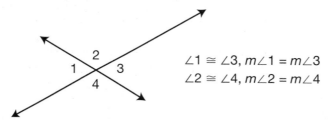

$$\angle1 \cong \angle3, m\angle1 = m\angle3$$
$$\angle2 \cong \angle4, m\angle2 = m\angle4$$

When two lines intersect and form four right angles, the lines are considered **perpendicular**.

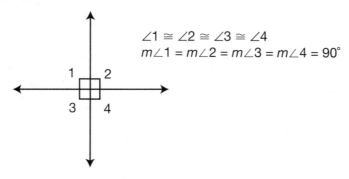

$$\angle1 \cong \angle2 \cong \angle3 \cong \angle4$$
$$m\angle1 = m\angle2 = m\angle3 = m\angle4 = 90°$$

Three-Lined Intersections

A **transversal line** intersects two or more lines, each at a different point. Because a transversal line crosses at least two other lines, eight or more angles are created. When a transversal intersects a pair of parallel lines, certain angles are always congruent or supplementary. Pairs of these angles have special names:

Corresponding angles are angles in corresponding positions.

Look for a distinctive F shaped figure.

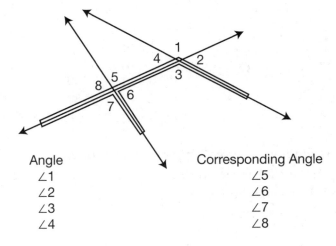

Angle	Corresponding Angle
∠1	∠5
∠2	∠6
∠3	∠7
∠4	∠8

When a transversal intersects a pair of parallel lines, **corresponding angles** are **congruent**.

Interior angles are angles inside a pair of crossed lines.

Look for a distinctive I shaped figure.

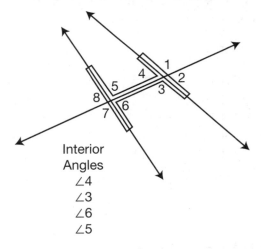

Interior
Angles
∠4
∠3
∠6
∠5

Same-side interior angles are interior angles on the same side of a transversal line.

Look for a distinctive C shaped figure.

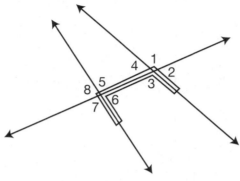

Same Side Interior Angles

| ∠3 | ∠6 |
| ∠4 | ∠5 |

When a transversal intersects a pair of parallel lines, **same-side interior angles** are **supplementary**.

Alternate interior angles are interior angles on opposite sides of a transversal line.

Look for a distinctive Z shaped figure.

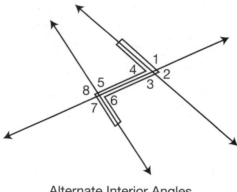

Alternate Interior Angles

| ∠4 | ∠6 |
| ∠3 | ∠5 |

When a transversal intersects a pair of parallel lines, **alternate interior angles** are **congruent**.

When a transversal is perpendicular to a pair of parallel lines, all eight angles are congruent.

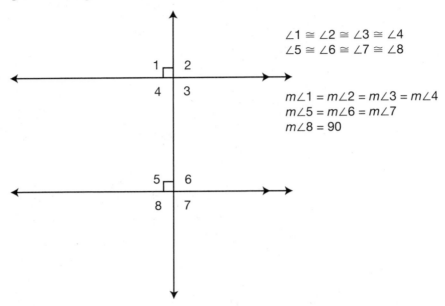

$\angle 1 \cong \angle 2 \cong \angle 3 \cong \angle 4$
$\angle 5 \cong \angle 6 \cong \angle 7 \cong \angle 8$

$m\angle 1 = m\angle 2 = m\angle 3 = m\angle 4$
$m\angle 5 = m\angle 6 = m\angle 7$
$m\angle 8 = 90$

There are also **exterior angles, same-side exterior angles,** and **alternate exterior angles.** They are positioned by the same common-sense rules as the interior angles.

Two lines are parallel if any one of the following statements is true:

1) A pair of alternate interior angles is congruent.
2) A pair of alternate exterior angles is congruent.
3) A pair of corresponding angles is congruent.
4) A pair of same-side interior angles is supplementary.

Set 11

Use the following diagram to answer questions 51 through 56.

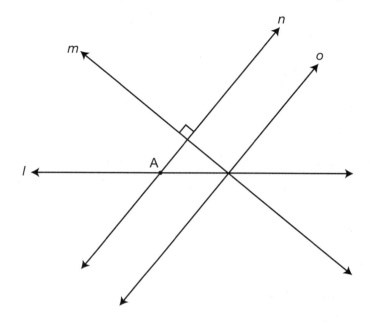

51. Which set of lines are transversals?
   a. *l, m, o*
   b. *o, m, n*
   c. *l, o, n*
   d. *l, m, n*

52. •A is
   a. between lines *l* and *n*.
   b. on lines *l* and *n*.
   c. on line *l*, but not line *n*.
   d. on line *n*, but not line *l*.

53. How many points do line *m* and line *l* share?
   a. 0
   b. 1
   c. 2
   d. infinite

54. Which lines are perpendicular?
   a. *n, m*
   b. *o, l*
   c. *l, n*
   d. *m, l*

55. How many lines can be drawn through •A that are perpendicular to line *l*?
   a. 0
   b. 1
   c. 10,000
   d. infinite

56. How many lines can be drawn through •A that are parallel to line *m*?
   a. 0
   b. 1
   c. 2
   d. infinite

Set 12

Use the following diagram to answer questions 57 through 61.

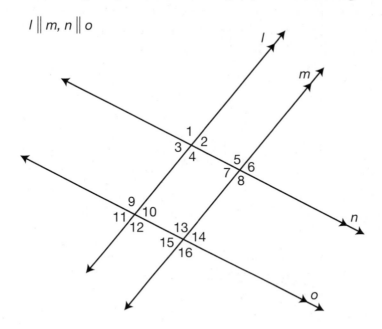

57. List all of the angles that are supplementary to ∠1.

58. State the special name for the following pair of angles: ∠13 and ∠16.

59. ∠5 is a corresponding angle to what other two angles?

60. In pairs, list all the alternate interior angles that are formed between parallel lines **l** and **m** and transversal **n**.

61. List all of the angles that are congruent to ∠10.

Set 13

Use the following diagram and the information below to determine if lines *o* and *p* are parallel. Place a checkmark (✓) beside statements that prove lines *o* and *p* are parallel; place an X beside statements that neither prove nor disprove that lines *o* and *p* are parallel.

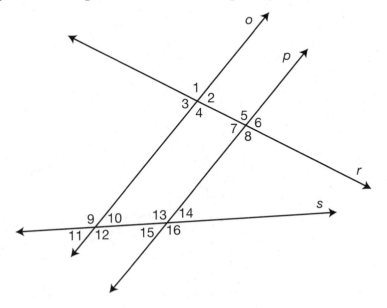

62. If ∠5 and ∠4 are congruent and equal, then _____.

63. If ∠1 and ∠2 are congruent and equal, then _____.

64. If ∠9 and ∠16 are congruent and equal, then _____.

65. If ∠12 and ∠15 are congruent and equal, then _____.

66. If ∠8 and ∠4 are congruent and equal, then _____.

Set 14

Circle the correct answer True or False. It may be helpful to draw the geometric situation before answering each question.

67. Angles formed by a transversal and two parallel lines are either supplementary or congruent.   **True** or **False**

68. When four rays extend from a single endpoint, adjacent angles are always supplementary.   **True** or **False**

69. Angles supplementary to the same angle or angles with the same measure are also equal in measure.   **True** or **False**

70. Adjacent angles that are also congruent are always right angles. **True** or **False**

71. Parallel and skew lines are coplanar.   **True** or **False**

72. Supplementary angles that are also congruent are right angles. **True** or **False**

73. If vertical angles are acute, the angle adjacent to them must be obtuse.   **True** or **False**

74. Vertical angles can be reflexive.   **True** or **False**

75. When two lines intersect, all four angles formed are never congruent to each other.   **True** or **False**

76. The sum of interior angles formed by a pair of parallel lines crossed by a transversal is always 360°.   **True** or **False**

77. The sum of exterior angles formed by a pair of parallel lines and a transversal is always 360°.   **True** or **False**

Answers

Set 11

51. **d.** In order to be a transversal, a line must cut across two other lines at different points. Line *o* crosses lines *m* and *l* at the same point; it is not a transversal.

52. **b.** When two lines intersect, they share a single point in space. That point is technically on both lines.

53. **b.** Lines are straight; they cannot backtrack or bend (if they could bend, they would be a curve, not a line). Consequently, when two lines intersect, they can share only one point.

54. **a.** When intersecting lines create right angles, they are perpendicular.

55. **b.** An infinite number of lines can pass through any given point in space—only one line can pass through a point and be perpendicular to an existing line. In this case, that point is on the line; however, this rule also applies to points that are not on the line.

56. **b.** Only one line can pass through a point and be parallel to an existing line.

Set 12

57. $\angle 1$ is supplementary to the two angles it is adjacent to, $\angle 2$ and $\angle 3$. All of the corresponding angles to $\angle 2$ and $\angle 3$ will also be supplementary to $\angle 1$. The following angles correspond to $\angle 2$: $\angle 6$, $\angle 10$, $\angle 14$. The following angles correspond to $\angle 3$: $\angle 7$, $\angle 11$, and $\angle 15$.

58. $\angle 13$ and $\angle 15$ are called **vertical angles** because they are the nonadjacent angles formed by two intersecting lines. Vertical angles are always congruent.

59. ∠5 corresponds to ∠1 since it is in the same position in respect to transversal **n**. ∠5 also corresponds to ∠13 since it is in the same position in respect to transversal **m**. (Although they are in the same relative position, ∠5 does not correspond to ∠9 since they share no common line.)

60. ∠2 and ∠7 are alternate interior angles since they are inside the parallel lines **l** and **m** and on opposite sides of transversal **n**. ∠4 and ∠5 are also alternate interior angles since they are inside the parallel lines **l** and **m** and on opposite sides of transversal **n**.

61. ∠10 is congruent to ∠11 since they are vertical; ∠14 and ∠2 since they are corresponding; ∠3 since it is congruent to ∠2; ∠15 since they are alternate interior; and ∠6 and ∠7 since those are corresponding and congruent to ∠2 and ∠3.

Set 13

62. ✓. Only three congruent angle pairs can prove a pair of lines cut by a transversal are parallel: alternate interior angles, alternate exterior angles, and corresponding angles. Angles 5 and 4 are alternate interior angles—notice the Z figure.

63. X. ∠1 and ∠2 are adjacent angles. Their measurements combined must equal 180°, but they do not determine parallel lines.

64. ✓. ∠9 and ∠16 are alternate exterior angles.

65. X. ∠12 and ∠15 are same side interior angles. Their congruence does not determine parallel lines. When same side interior angles are supplementary, then the lines are parallel.

66. ✓. ∠8 and ∠4 are corresponding angles.

Set 14

67. True. The angles of a pair of parallel lines cut by a transversal are always either supplementary or congruent, meaning their measurements either add up to 180°, or they are the same measure.

68. False. If the four rays made two pairs of opposite rays, then this statement would be true; however, any four rays extending from a single point do not have to line up into a pair of straight lines; and without a pair of straight lines there are no supplementary angle pairs.

69. True.

70. False. Adjacent angles do not always form straight lines; to be adjacent, angles need to share a vertex, a side, and no interior points. However, adjacent angles that do form a straight line are always right angles.

71. False. Parallel lines are coplanar; skew lines are not.

72. True. A pair of supplementary angles must measure 180°. If the pair is also congruent, they must measure 90° each. An angle that measures 90° is a right angle.

73. True. When two lines intersect, they create four angles. The two angles opposite each other are congruent. Adjacent angles are supplementary. If vertical angles are acute, angles adjacent to them must be obtuse in order to measure 180°.

74. False. Vertical angles cannot be equal to or more than 180°; otherwise, they could not form supplementary angle pairs with their adjacent angle.

75. False. Perpendicular lines form all right angles.

76. True. Adjacent interior angles form supplementary pairs; their joint measurement is 180°. Two sets of adjacent interior angles must equal 360°.

77. True. Two sets of adjacent exterior angles must equal 360°.

4

Measuring Angles

Had enough of angles? You haven't even begun! You named angles and determined their congruence or incongruence when two lines cross. In this chapter, you will actually measure angles using an instrument called the **protractor**, a special ruler used to measure the degrees of angles.

How to Measure an Angle Using a Protractor

Most protractors have two scales along their arc. The lower scale, starting with zero on the right, is used to **measure angles** that open in a counter-clockwise arc. To measure a counter-clockwise angle, place the vertex in the circular opening at the base of the protractor and extend the bottom ray through the 0° mark out to the right. In Figure 4.1, the counter clockwise angle drawn is 20°:

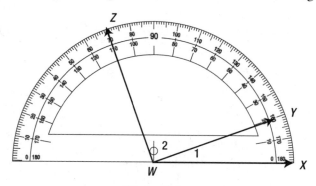

Figure 4.1

The second scale along the top, which begins with zero on the left, measures angles that open in a clockwise arc. To measure a clockwise angle, place the vertex in the circular opening at the base of the protractor and extend the bottom ray through the 0° mark out to the left. In Figure 4.2, the clockwise angle drawn is 120°:

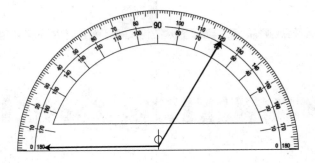

Figure 4.2

How to Draw an Angle Using a Protractor

To draw an angle, first draw a ray. The ray's end point becomes the angle's vertex. Position the protractor as if you were measuring an angle. Choose your scale and make a mark on the page at the desired measurement. Remove the protractor and connect the mark you made to the vertex with a straight edge. Voilà, you have an angle.

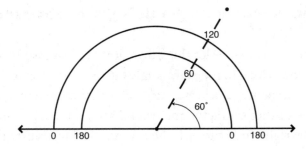

Adjacent Angles

Adjacent angles share a vertex, a side, and no interior points; they are angles that lie side-by-side.

Because adjacent angles share a single vertex point, adjacent angles can be added together to make larger angles. This technique will be particularly useful when working with **complementary** and **supplementary** angles in Chapter 5.

Angle and Line Segment Bisectors

The prefix "bi" means "two," as in *bicycle*, which has two wheels, or bicentennial, which is a two-hundredth anniversary. In geometry, a **bisector** is any ray or line segment that divides an angle or another line segment into two congruent and equal parts. Sometimes you will read that a ray "**bisects** an angle" and other times a ray may be referred to as an "angle bisector," but either way, it means that the ray divides the bisected angle into two smaller, equal angles. In the following illustration, we can tell that ∠CAK is bisected by ray AR, since ∠CAR ≅ ∠RAK. We know these two angles are congruent because they both have double tick marks in them—when angles have the same amount of tick marks, it means that they are congruent.

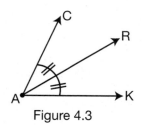
Figure 4.3

Set 15

Using the diagram below, measure each angle.

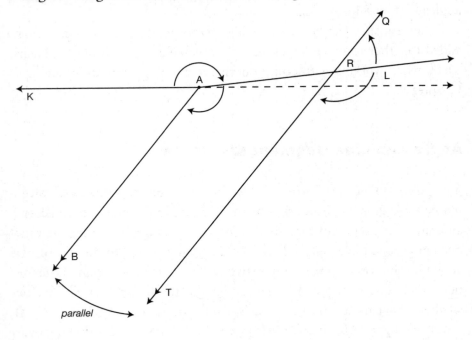

78. ∠LRQ

79. ∠ART

80. ∠KAL

81. ∠KAB

82. ∠LAB

Set 16

Using a protractor, draw a figure starting with question 83. Complete question 87 by using your drawn figure.

83. Draw $\overrightarrow{EC}$.

84. $\overrightarrow{ED}$ rotates 43° counterclockwise (left) from $\overrightarrow{EC}$. Draw $\overrightarrow{ED}$.

85. $\overrightarrow{EF}$ rotates 90° counterclockwise from $\overrightarrow{ED}$. Draw $\overrightarrow{EF}$.

86. $\overrightarrow{EG}$ and $\overrightarrow{EF}$ are opposite rays. Draw $\overrightarrow{EG}$.

87. What is the measurement of ∠GEC?

Set 17

Choose the best answer.

88. Consider any two angles, ∠ROT and ∠POT. ∠ROT and ∠POT are
    a. supplementary angles.
    b. right angles.
    c. congruent angles.
    d. adjacent angles.
    e. No determination can be made.

89. When adjacent angles RXZ and ZXA are added, they make
    a. ∠RXA.
    b. ∠XZ.
    c. ∠XRA.
    d. ∠ARX.
    e. No determination can be made.

90. Adjacent angles EBA and EBC make ∠ABC. ∠ABC measures 132°. ∠EBA measures 81°. ∠EBC must measure
    a. 213°.
    b. 61°.
    c. 51°.
    d. 48°.
    e. No determination can be made.

91. ∠SVT and ∠UVT are adjacent supplementary angles. ∠SVT measures 53°. ∠UVT must measure
    a. 180°.
    b. 233°.
    c. 133°.
    d. 127°.
    e. No determination can be made.

92. ∠AOE is a straight angle. ∠BOE is a right angle. ∠AOB is
    a. a reflexive angle.
    b. an acute angle.
    c. an obtuse angle.
    d. a right angle.
    e. No determination can be made.

Set 18

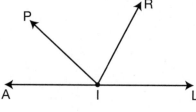

Figure 4.4

Answer the following questions using the figure above.

93. If $\overrightarrow{IR}$ bisects ∠PIL and the measurement of ∠RIL = 68°, then what is the measurement of ∠PIL?

94. Using the information from question 93, if $\overleftrightarrow{AL}$ is a straight angle, what is the measurement of ∠PIA?

Answers

Set 15

78. $m\angle LRQ = 45$

79. $m\angle ART = 45$

80. $m\angle KAL = 174$

81. $m\angle KAB = 51$

82. $m\angle LAB = 135$

Set 16

83.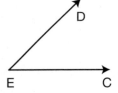

84.

85.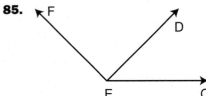

86.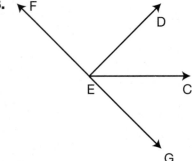

87. We know that FG is a straight angle. Since ∠FED measures 90°, that means that ∠GED also equals 90°. Since ∠DEC + ∠CEG equals 90°, use $m∠DEC = 43°$ to determine that ∠CEG = 47°.

Set 17

88. **e.** ∠ROT and ∠POT share a vertex point and one angle side. However, it cannot be determined that they do not share any interior points, that they form a straight line, that they form a right angle, or that they are the same shape and size. The answer must be choice **e.**

89. **a.** When angles are added together to make larger angles, the vertex always remains the same. Choices **c** and **d** move the vertex point to •R; consequently, they are incorrect. Choice **b** does not name the vertex at all, so it is also incorrect. Choice **e** is incorrect because we are given that the angles are adjacent; we know they share side XZ; and we know they do not share sides XR and XA. This is enough information to determine they make ∠RXA.

90. **c.** EQUATION:
$m∠ABC - m∠EBA = m∠EBC$
$132° - 81° = 51°$

91. **d.** EQUATION:
$m∠SVT + m∠UVT = 180°$
$53° + m∠UVT = 180°$
$m∠UVT = 127°$

92. **d.** Draw this particular problem out; any which way you draw it, ∠AOB and ∠BOE are supplementary. 90° subtracted from 180° equals 90°. ∠AOB is a right angle.

Set 18

93. Since ∠PIL is bisected by ∠RIL, we know that ∠PIL is divided into two equal angles. Therefore, since $m\angle RIL = 68°$, it follows that $m\angle PIL = 68°$.

94. Since $\overleftrightarrow{AL}$ is a straight angle, then $m\angle AIL = 180°$. Subtract ∠PIL from 180° in order to find $m\angle PIA$. $180° - 136° = 44°$. Therefore, $m\angle PIA = 44°$.

5

Pairs of Angles

Well done! Good job! Excellent work! You have mastered the use of protractors. You can now move into an entire chapter dedicated to complements and supplements. Perhaps the three most useful angle pairs to know in geometry are complementary, supplementary, and vertical angle pairs.

Complementary Angles

When two adjacent or nonadjacent angles have a total measurement of 90°
they are **complementary angles**.

You can probably see that these are similar to supplementary angles,
except complementary angles add to 90° instead of to 180°. The following
two illustrations show examples of **adjacent complementary angles** and
non-adjacent complementary angles.

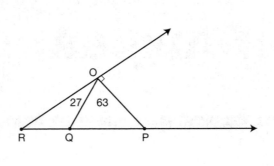

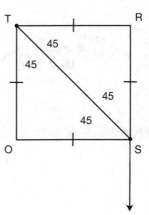

$m\angle ROQ + m\angle QOP = 90°$
$\angle ROQ$ and $\angle QOP$ are
adjacent complementary angles

$m\angle OTS + m\angle TSO = 90°$
$\angle OTS$ and $\angle TSO$ are
nonadjacent complementary
angles

Supplementary Angles

When two adjacent or nonadjacent angles have a total measure of 180° they
are **supplementary angles**.

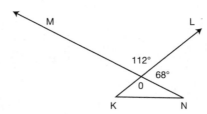

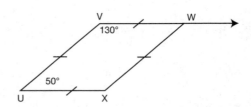

$m\angle MOL + m\angle LON = 180°$
$\angle MOL$ and $\angle LON$ are
adjacent supplementary angles

$m\angle XUV + m\angle UVW = 180°$
$\angle XUV$ and $\angle UVW$ are nonadjacent supplementary angles

Vertical Angles

When two straight lines intersect, the nonadjacent pairs of angles formed (the angles that are opposite from each other), are called **vertical angles**. Vertical angles are always congruent and have equal angle measures.

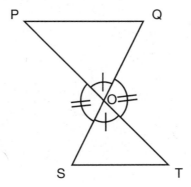

∠POT and ∠QOS are straight angles
∠POQ ≅ ∠SOT $m∠POQ = m∠SOT$
∠POQ and ∠SOT are vertical angles
∠POS ≅ ∠QOT $m∠POS = m∠QOT$
∠POS and ∠QOT are vertical angles

Other Angles That Measure 180°

When three line segments intersect at three unique points to make a closed figure, they form a triangle. The interior angles of the triangle are the angles inside the closed figure. The sum of a triangle's three interior angles is always 180°.

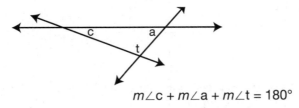

$$m∠c + m∠a + m∠t = 180°$$

Set 19

Choose the best answer for questions 95 through 99 based on the figure below.

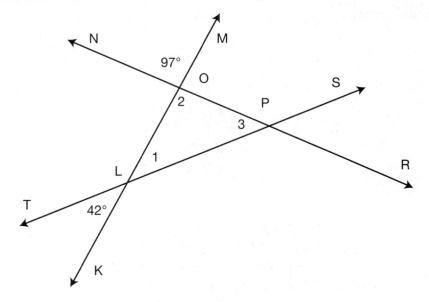

95. Name the angle vertical to ∠NOM.
   a. ∠NOL
   b. ∠KLP
   c. ∠LOP
   d. ∠MOP

96. Name the angle vertical to ∠TLK.
   a. ∠MOR
   b. ∠NOK
   c. ∠KLT
   d. ∠MLS

97. Which two angles are supplementary to ∠NOM.
   a. ∠MOR and ∠NOK
   b. ∠SPR and ∠TPR
   c. ∠NOL and ∠LOP
   d. ∠TLK and ∠KLS

98. ∠1, ∠2, and ∠3 respectively measure
   a. 90°, 40°, 140°.
   b. 139°, 41°, 97°.
   c. 42°, 97°, 41°.
   d. 41°, 42°, 83°.

99. The measurement of ∠OPS is
   a. 139°.
   b. 83°.
   c. 42°.
   d. 41°.

Set 20

Choose the best answer.

100. If ∠LKN and ∠NOP are complementary angles,
   a. they are both acute.
   b. they must both measure 45°.
   c. they are both obtuse.
   d. one is acute and the other is obtuse.
   e. No determination can be made.

101. If ∠KAT and ∠GIF are supplementary angles,
   a. they are both acute.
   b. they must both measure 90°.
   c. they are both obtuse.
   d. one is acute and the other is obtuse.
   e. No determination can be made.

102. If ∠DEF and ∠IPN are congruent, they are
   a. complementary angles.
   b. supplementary angles.
   c. right angles.
   d. adjacent angles.
   e. No determination can be made.

103. If ∠ABE and ∠GIJ are congruent supplementary angles, they are
   a. acute angles.
   b. obtuse angles.
   c. right angles.
   d. adjacent angles.
   e. No determination can be made.

104. If ∠EDF and ∠HIJ are supplementary angles, and ∠SUV and ∠EDF are also supplementary angles, then ∠HIJ and ∠SUV are
   a. acute angles.
   b. obtuse angles.
   c. right angles.
   d. congruent angles.
   e. No determination can be made.

Set 21

Answer questions 105–108 based on your knowledge of angles and the figure below.

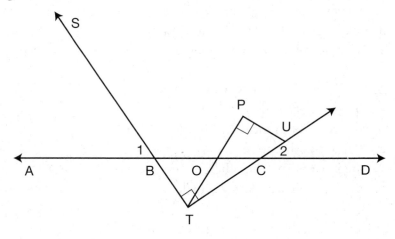

105. If $m∠ABT = 125°$, then what is the $m∠TBO$?

106. If $m∠BTO = 66°$, then what is the $m∠OTC$?

107. Use the information given in questions **105** and **106** and your answers calculated to determine the $m\angle$BOT.

108. Using your answer from question **106**, determine the $m\angle$PUT.

Set 22

State the relationship or sum of the angles given based on the figure below. If a relationship cannot be determined, then state, "They cannot be determined."

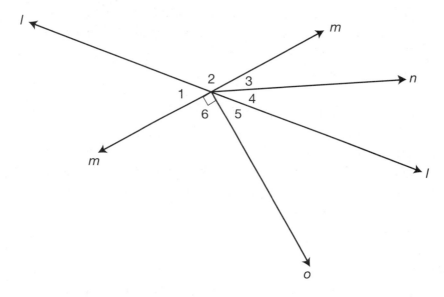

109. Measurement of $\angle2$ and the measuresment of $\angle6$ plus $\angle5$.

110. $\angle1$ and $\angle3$.

111. $\angle1$ and $\angle2$.

112. The sum of $\angle5$, $\angle4$, and $\angle3$.

113. $\angle6$ and $\angle2$.

114. The sum of $\angle1$, $\angle6$, and $\angle5$.

Answers

Set 19

95. **c.** ∠NOM and ∠LOP are opposite angles formed by intersecting lines NR and MK; thus, they are vertical angles.

96. **d.** ∠TLK and ∠MLS are opposite angles formed by intersecting lines TS and MK; thus, they are vertical angles.

97. **a.** ∠MOR and ∠NOK are both adjacent to ∠NOM along two different lines. The measurement of each angle added to the measurement of ∠NOM equals that of a straight line, or 180°. Each of the other answer choices is supplementary to each other, but not to ∠NOM.

98. **c.** ∠1 is the vertical angle to ∠TLK, which is given. ∠2 is the vertical pair to ∠NOM, which is also given. Since vertical angles are congruent, ∠1 and ∠2 measure 42° and 97°, respectively. To find the measure of ∠3, subtract the sum of ∠1 and ∠2 from 180° (the sum of the measurement of a triangle's interior angles):
$180° - (42° + 97°) = m\angle 3$
$41° = m\angle 3$

99. **a.** There are two ways to find the measurement of exterior angle OPS. The first method subtracts the measurement of ∠3 from 180°. The second method adds the measurements of ∠1 and ∠2 together because the measurement of an exterior angle equals the sum of the two nonadjacent interior angles. ∠OPS measures 139°.

Set 20

100. **a.** The sum of any two complementary angles must equal 90°. Any angle less than 90° is acute. It only makes sense that the measurement of two acute angles could add to 90°. Choice **b** assumes both angles are also congruent; however, that information is not given. If the measurement of one obtuse angle equals more than 90°, then two obtuse angles could not possibly measure exactly 90° together. Choices **c** and **d** are incorrect.

101. **e.** Unlike the question above, where every complementary angle must also be acute, supplementary angles can be acute, right, or obtuse. If an angle is obtuse, its supplement is acute. If an angle is right, its supplement is also right. Two obtuse angles can never be a supplementary pair, and two acute angles can never be a supplementary pair. Without more information, this question cannot be determined.

102. **e.** Complementary angles that are also congruent measure 45° each. Supplementary angles that are also congruent measure 90° each. Without more information, this question cannot be determined.

103. **c.** Congruent supplementary angles always measure 90° each:

$m\angle ABE = x$

$m\angle GIJ = x$

$m\angle ABE + m\angle GIJ = 180°$; replace each angle with its measure:

$x + x = 180°$

$2x = 180°$; divide each side by 2:

$x = 90°$

Any 90° angle is a right angle.

104. **d.** When two angles are supplementary to the same angle, they are congruent to each other:

$m\angle EDF + m\angle HIJ = 180°$

$m\angle EDF + m\angle SUV = 180°$

$m\angle EDF + m\angle HIJ = m\angle SUV + m\angle EDF$; subtract $m\angle EDF$ from each side:

$m\angle HIJ = m\angle SUV$

Set 21

105. Since $\angle ABO$ is a straight angle, you know that $m\angle ABO = 180°$. Therefore $m\angle ABT + m\angle TBO = 180°$. Since $m\angle ABT = 125°$, then $\angle TBO$ must equal 55°.

106. Since $\angle BTC$ is a right angle, you know that $m\angle BTC = 90°$. Therefore $m\angle BTO + m\angle OTC = 90°$. Since $m\angle BTO = 66°$, then $\angle TBO$ must equal 24°.

107. The interior angles of ∆BOT sum to 180° and since you know that $m\angle TBO = 55°$, and $m\angle BTO = 66°$, then you need to solve for $m\angle BOT$ in the equation = 55° + 66° + $m\angle BOT$ = 180°. Using algebra to solve the equation, the $m\angle BOT = 59°$.

108. The interior angles of ∆PUT sum to 180° and since you know that $m\angle PTU = 24°$ (from question **106**), and $m\angle TPU = 90°$, then you need to solve for $m\angle PUT$ in the equation = 24° + 90° + $m\angle PUT$ = 180°. Using algebra to solve the equation, the $m\angle PUT = 64°$.

Set 22

109. **Equal.** Together $\angle 5$ and $\angle 6$ form the vertical angle pair to $\angle 2$. Consequently, the angles are congruent and their measurements are equal.

110. **They cannot be determined.** $\angle 1$ and $\angle 3$ may look like vertical angles, but do not be deceived. Vertical angle pairs are formed when lines intersect. The vertical angle to $\angle 1$ is the full angle that is opposite and between lines m and l.

111. **Adjacent supplementary angles.** $\angle 1$ and $\angle 2$ share a side, a vertex and no interior points; they are adjacent. The sum of their measures must equal 180° because they form a straight line; thus they are supplementary.

112. **90°.** $\angle 6$, $\angle 5$, $\angle 4$, and $\angle 3$ are on a straight line. All together, they measure 180°. If $\angle 6$ is a right angle, it equals 90°. The remaining three angles must equal 180° minus 90°, or 90°.

113. **They cannot be determined.** $\angle 6$ and $\angle 2$ may look like vertical angles, but vertical pairs are formed when lines intersect. The vertical angle to $\angle 2$ is the full angle that is opposite and between lines m and l.

114. 180°.

6

Types of Triangles

Mathematicians have an old joke about angles being very extroverted. How so? Because they are always open! The two rays of an angle extend out in different directions and continue on forever. Who are the introverts in mathematics? If you connect three or more line segments end-to-end you will have a very, shy *closed-figure*, called a polygon.

By definition, polygons are closed geometric figures that have three or more *straight* line segments as sides. Polygons cannot contain arcs or curves, as illustrated in the figures below.

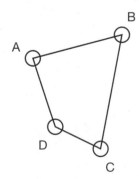

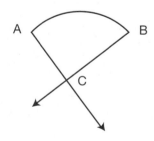

Polygon

- made of all line segments
- each line segment exclusively meets the end of another line segment
- all line segments make a closed figure

NOT a Polygon

- $\overset{\frown}{AB}$ is not a line segment
- C is not an endpoint
- Figure ABC is not a closed figure ($\overrightarrow{AC}$ and $\overrightarrow{BC}$ extend infinitely)

Triangles

The simplest polygon is the triangle. It has the fewest sides and angles that a polygon can have.

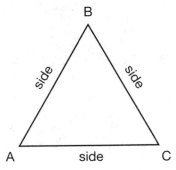

ΔABC

Sides: $\overline{AB}$, $\overline{BC}$ and $\overline{CA}$

Vertices: ∠ABC, ∠BCA, and ∠CAB

Triangles can be classified by their angles or by their sides. When looking at the congruence or incongruence of triangle sides, all triangles can be grouped into one of three special categories.

Naming Triangles by Their Sides

1. Scalene triangles have *no congruent sides and no congruent angles.*

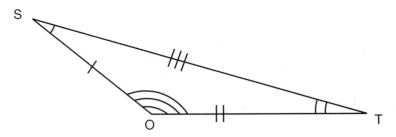

- ΔSOT
- $\overline{ST} \neq \overline{TO} \neq \overline{OS}$
- ∠STO ≇ ∠TOS ≇ ∠OST
- ΔSOT is scalene

2. **Isosceles triangles** have two congruent sides and two congruent angles. The two congruent angles are called the **base angles** and the unique angle is called the **vertex**. The two congruent sides are called the **legs** and the unique side is called the **base**.

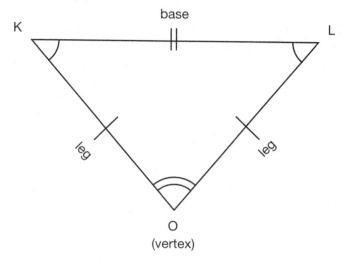

- ΔKLO
- $\overline{KO} \cong \overline{LO}$
- ∠LKO ≅ ∠KLO
- ΔKLO is isosceles

3. **Equilateral triangles** have three congruent sides and three congruent angles. The three congruent angles will always each measure 60°. (This is true because the 180° of the triangles interior angles divided by three angles equals 60°.)

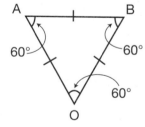

- ΔABO
- $\overline{AB} \cong \overline{BO} \cong \overline{OA}$
- ∠ABO ≅ ∠BOA ≅ ∠OAB
- ΔABO is equilateral

Naming Triangles by Their Angles

You just learned how triangles can be classified by their sides and now you will learn how to classify triangles by their angles. Triangles can be classified by the *size* of their angles or by the *congruence* of their angles. First let's look at how triangles are classified by the *size* of their angles.

1. **Acute triangles** have three acute angles; none of the angles are 90° or greater than 90°. ΔEOF below is an acute triangle.

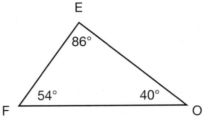

Acute Triangle EOF $m\angle$EOF, $m\angle$OFE
                                              and $m\angle$FEO < 90°

2. **Obtuse triangles** have one angle that is obtuse, or greater than 90°, and two acute angles. (Remember, it is impossible for a triangle to have more than one obtuse angle because the sum of the interior angles of a triangle always equals 180°.) The following ΔMOL is an obtuse triangle.

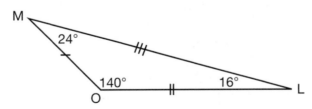

- Obtuse Triangle LMO
- $m\angle$LOM > 90°
- $m\angle$OLM and $m\angle$LMO < 90°

3. **Right triangles** have one right angle that measures 90° and two acute angles. In right triangles, the right angle is always opposite the longest side of the triangle, which is called the **hypotenuse**. The other two sides, which are the sides of the right angle, are called the **legs**. (Remember, it is impossible for a triangle to have two right angles because the sum of the interior angles of a triangle always equals 180°.) The following ΔTOS is a right triangle.

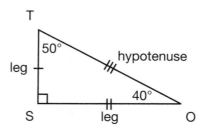

- Right Triangle TOS
- $m\angle TSO = 90°$
- $m\angle TOS$ and $m\angle STO < 90°$

We've just looked at how to classify triangles by the congruence of their sides and by the size of their angles. Lastly, let's focus on how to classify triangles by the congruence of their angles. Luckily, you already had a preview of this in the beginning of this chapter!

1. **Scalene triangles** have no congruent angles (and no congruent sides).
2. **Isosceles triangles** have two congruent angles (and two congruent sides).
3. **Equilateral triangles** have three congruent angles (and three congruent sides). The three congruent angles always each equal 60°.

One thing to keep in mind is that acute, obtuse, and right triangles can also be classified as scalene, isosceles, or equilateral. For example, ΔORQ in the following figure is both an isosceles triangle and a right triangle: it has a vertex that is 90° and two congruent base angles that are each 45°.

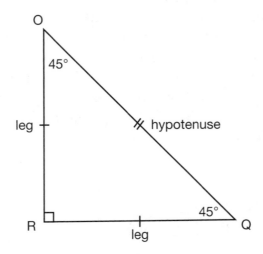

- Isosceles Triangle ORQ is also Right Triangle ORQ
- $m\angle ORQ = 90°$
- $m\angle ROQ = m\angle RQO < 90°$

Another example of how a triangle can fall into two classification groups is illustrated in the following triangle, ΔKJO. $\angle J$ is an obtuse angle and $\angle O = \angle K$, which makes ΔKJO an isosceles triangle and an obtuse triangle. It can be said that ΔKJO is an *obtuse isosceles triangle*.

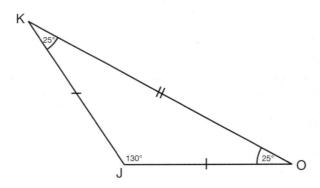

- Obtuse Isosceles Triangle JKO
- $m\angle OJK > 90°$
- $m\angle JKO$ and $m\angle KOJ < 90°$

Relationships between Sides and Angles in Triangles

There is a special relationship between the lengths of the sides of a triangle and the relative measures of each side's opposite angle. *The longest side of a triangle is always opposite the largest angle, the shortest side is always opposite the smallest angle, and the middle-length side is always opposite the middle-sized angle.* In the case that two angles are equal, then the two sides that are opposite them will also be equal, and vice versa. Therefore, it is impossible for the largest angle of a triangle to be opposite anything but the largest side. For example, look at the following triangle, $\triangle PDX$. Since $\overline{PX}$ is the largest side, $\angle D$ will be the largest; since $\overline{PD}$ is the shortest side, $\angle X$ will be the shortest; and lastly, since $\overline{DX}$ is the middle side, $\angle P$ will be the middle in size.

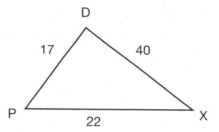

We can use the angles in $\triangle MAE$ below to get similar information. Since $\angle M = \angle E$, then you can be certain that $\overline{MA} = \overline{EA}$.

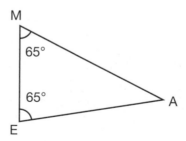

The last fact to keep in mind about triangles is that sum of any two sides of a triangle must be greater than the length of the remaining third side. For example, if two sides of a triangle are 8 and 10 the third side must be shorter than 18. If it were longer than 18, the other two sides would not be able to connect to the side's endpoints. This is illustrated below in $\triangle AUG$; notice that when $\overline{AG} \geq 18$, it is too long for the triangle to close.

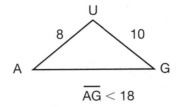 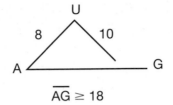

$\overline{AG} < 18$ $\overline{AG} \geq 18$

Now you can practice classifying triangles!

Set 23

State the name of the triangle based on the measures given. Remember that many triangles can have more than one classification. If the information describes a figure that cannot be a triangle, write, "Cannot be a triangle." It will be helpful to draw the information you are given.

115. ΔBDE, where $\overline{BD}$ = 17, $\overline{BE}$ = 22, $\angle D$ = 47°, and $\angle B$ = 47°.

116. ΔQRS, where $m\angle R$ = 94°, $m\angle Q$ = 22° and $m\angle S$ = 90°.

117. ΔWXY, where $\overline{WX}$ = 10, $\overline{XY}$ = 10, $\overline{YW}$ = 10, and $m\angle X$ = 90°.

118. ΔPQR, where $m\angle P$ = 31° and $m\angle R$ = 89°.

119. ΔABD, where $\overline{AB}$ = 72, $\overline{AD}$ = 72 and $m\angle A$ = 90°.

120. ΔTAR, where $\overline{TA}$ = 20, $\overline{AR}$ = 24, and $\overline{TR}$ = 44.

121. ΔDEZ, where $m\angle 1$ = 60° and $m\angle 2$ = 60°.

122. ΔCHI, where $m\angle 1$ = 30°, $m\angle 2$ = 60° and $m\angle 3$ = 90°.

123. ΔJMR, where $m\angle 1$ = 5°, $m\angle 2$ = 120° and $m\angle 3$ = 67°.

124. ΔKLM, where $\overline{KL}$ = $\overline{LM}$ = $\overline{MK}$.

Set 24

Fill in the blanks based on your knowledge of triangles and angles.

125. In right triangle ABC, if ∠C measures 31° then the remaining acute angle must measure _____.

126. In scalene triangle QRS, if ∠R measures 134° and ∠Q measures 16°, then ∠S measures _____.

127. In isosceles triangle TUV, if vertex ∠T is supplementary to an angle in an equilateral triangle, then base ∠U measures _____.

128. In obtuse isosceles triangle EFG, if the base ∠F measures 12°, then the vertex ∠E measures _____.

129. In acute triangle ABC, if ∠B measures 45°, can ∠C measure 30°? _____.

Set 25

Choose the best answer.

130. Which of the following sets of interior angle measurements would describe an acute isosceles triangle?
   a. 90°, 45°, 45°
   b. 80°, 60°, 60°
   c. 60°, 60°, 60°
   d. 60°, 50°, 50°

131. Which of the following sets of interior angle measurements would describe an obtuse isosceles triangle?
   a. 90°, 45°, 45°
   b. 90°, 90°, 90°
   c. 100°, 50°, 50°
   d. 120°, 30°, 30°

132. Which of the following angle measurements **could not** describe an interior angle of a right angle?

    a. 30°

    b. 60°

    c. 90°

    d. 100°

133. If △JNM is equilateral and equiangular, which condition would not exist?

    a. $\overline{JN} = \overline{MN}$

    b. $\overline{JM} \cong \overline{JN}$

    c. $m\angle N = m\angle J$

    d. $m\angle M = \overline{NM}$

134. In isosceles △ABC, if vertex ∠A is twice the measure of base ∠B, then ∠C measures

    a. 30°.

    b. 33°.

    c. 45°.

    d. 90°.

Set 26

Using the obtuse triangle diagram below, determine which pair of given angles has a greater measurement in questions 135 through 138. Note: $m\angle 2 = 111°$.

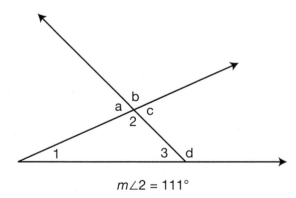

$m\angle 2 = 111°$

135. ∠3 or ∠b

136. ∠3 or ∠d

137. ∠a or ∠b

138. ∠1 or ∠c

For questions 139 through 141, use ΔNPR, which is drawn to scale.

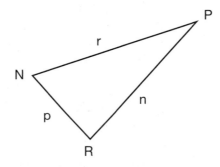

139. Given that ΔNPR is drawn to scale, which angle is the smallest and why?

140. Given that ΔNPR is drawn to scale, which side is the longest and why?

141. If side r = 17mm and side p = 8mm, what is the largest whole-numbered side length that could be possible for side n?

Answers

Set 23

115. **Cannot be a triangle.** Since $m\angle D = m\angle B = 47°$, then this is an isosceles triangle with a vertex at $\angle E$ that equals 86° (180° − (47°) · (2) = 86°). Since $\angle E$ is the largest angle, it must be opposite the largest side, but it is opposite the smallest side, 17. Therefore, this cannot be a triangle.

116. **Cannot be a triangle.** Any triangle can have one right angle or one obtuse angle, not both. "Triangle" QRS claims to have a right angle and an obtuse angle.

117. **Cannot be a triangle.** "Triangle" WXY claims to be equilateral and right; however, an equilateral triangle also has three congruent interior angles, and no triangle can have three right angles.

118. **Acute scalene triangle PQR.** Subtract from 180° the sum of $\angle P$ and $\angle R$. $\angle Q$ measures 60°. All three angles are acute, and all three angles are different. ΔPQR is acute scalene.

119. **Isosceles right triangle ABD.** $\angle A$ is a right angle and sides AB = AD.

120. **Cannot be a triangle.** In triangles, the sum of any two sides of a triangle must be greater than the length of the remaining third side. In this case 20 + 24 = 44, which is the length of the third side. This means the sum of two of the sides of the triangle is *equal to* and not *greater than* the remaining side, so ΔTAR cannot exist.

121. **Acute equilateral triangle DEZ.** Subtract from 180° the sum of $\angle 1$ and $\angle 2$. $\angle 3$, like $\angle 1$ and $\angle 2$, measures 60°. An equiangular triangle is an equilateral triangle, and both are always acute.

122. **Scalene right triangle CHI.** $\angle 3$ is a right angle; $\angle 1$ and $\angle 2$ are acute; and all three sides have different lengths.

123. **Cannot be a triangle.** Add the measurements of each angle together. The sum exceeds 180° so ΔJMR cannot exist.

124. **Acute equilateral triangle KLM.** When all three codes are congruent, you have an equilateral triangle.

Set 24

125. **59°.** $180° - (m\angle C + m\angle A) = m\angle B$. $180° - 121° = m\angle B$. $59° = m\angle B$

126. **30°.** $180° - (m\angle R + m\angle Q) = m\angle S$. $180° - 150° = m\angle S$. $30° = m\angle S$

127. **30°.** Step One: $180° - 60° = m\angle T$. $120° = m\angle T$. Step Two: $180° - m\angle T = m\angle U + m\angle V$. $180° - 120° = m\angle U + m\angle V$. $60° = m\angle U + m\angle V$. Step Three: 60° shared by two congruent base angles equals two 30° angles.

128. **156°.** $180° - (m\angle F + m\angle G) = m\angle E$. $180° - 24° = m\angle E$. $156° = m\angle E$

129. **No.** The sum of the measurements of $\angle B$ and $\angle C$ equals 75°. Subtract 75° from 180°, and $\angle A$ measures 105°. $\triangle ABC$ cannot be acute if any of its interior angles measure 90° or more.

Set 25

130. **c.** Choice **a** is not an acute triangle because it has one right angle. In choice **b**, the sum of interior angle measures exceeds 180°. Choice **d** suffers the reverse problem; its sum does not make 180°. Though choice **c** describes an equilateral triangle; it also describes an isosceles triangle.

131. **d.** Choice **a** is not an obtuse triangle; it is a right triangle. In choice **b** and choice **c** the sum of the interior angle measures exceeds 180°.

132. **d.** A right triangle has a right angle and two acute angles; it does not have any obtuse angles.

133. **d.** Angles and sides are measured in different units. 60 inches is not the same as 60°.

134. **c.** Let $m\angle A = 2x$, $m\angle B = x$ and $m\angle C = x$. $2x + x + x = 180°$. $4x = 180°$. $x = 45°$.

Set 26

135. ∠**b.** ∠b is the vertical angle to obtuse ∠2, which means ∠b is also obtuse. The measurements of ∠2 and ∠b both exceed the measure of ∠3.

136. ∠**d.** If ∠3 is acute, its supplement must be obtuse.

137. ∠**b.** ∠b is vertical to obtuse angle 2, which means ∠b is also obtuse. The supplement to an obtuse angle is always acute.

138. ∠**c.** The measurement of an exterior angle equals the sum of nonadjacent interior angles. Thus, the measurement of ∠c equals the measurement of ∠1 plus the measurement of ∠3. It only makes sense that the measurement of ∠c is greater than the measurement of ∠1 all by itself.

139. Since side p is the shortest side, ∠P, the angle opposite it, will be the smallest.

140. Since ∠R is the largest angle, the side that is opposite it, side r, will be the longest.

141. If side r = 17mm and side p = 8mm, their sum of 25mm must be larger than the length of the remaining side, n. Therefore, the largest whole-numbered side length for side n is 24mm.

7

Congruent Triangles

When you look in a regular bathroom mirror you will see your reflection. You will be the same shape and same size. In geometry, when figures are exact duplicates, we say they are congruent. When you look at a 3 × 5 photograph of yourself, you will also look the same, but much smaller. That 3 × 5 photo could be enlarged to fit on a billboard, and then it would be a much larger version of yourself.

In geometry, figures that are identical in all aspects of their appearance *other than size*, such as the 3 × 5 photo of you and the enlarged billboard image of you, are called **similar**. Lastly, figures that are not alike at all, like a picture of you and a picture of your dog, are called **dissimilar**.

Some Congruency Basics

When items are **congruent** it means they are identical. In order to indicate congruence, the following symbol is used: ≅ . Therefore, ΔA ≅ ΔB, is read, "triangle A is congruent to triangle B."

Congruent line segments are line segments that have the same length, regardless of how they are oriented on paper.

Congruent angles are angles that have the same measurement in degrees. Since they do not need to open or face the same direction, sometimes they can be difficult to identify. Congruent angles are shown with a small arc and an identical number of hash marks. In the following illustration, ∠XYZ ≅ ∠ABC even though they are facing different directions.

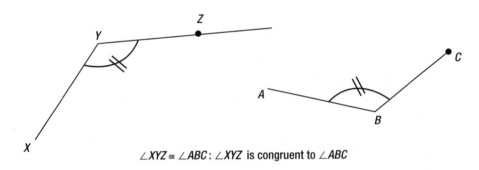

∠XYZ ≅ ∠ABC: ∠XYZ is congruent to ∠ABC

Naming congruent figures accurately takes careful consideration. Looking at the following figure, you might remember that it does not matter if we refer to ΔABC as ΔBCA or as ΔCAB since *order* and *direction* are unimportant when *naming* a single triangle. However, when discussing congruent figures, the order of vertexes from one figure must correspond with the congruent vertexes from the other figure. For example, if we were to use ΔABC, then we would need to start the name of the second triangle with the smallest angle Q, and follow it with the largest angle R, and end with the middle angle S. Therefore, ΔABC ≅ ΔQRS. It would be *incorrect* to write ΔABC ≅ ΔQSR because that implies that <B is congruent to <S, but these angles are not congruent.

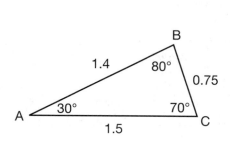
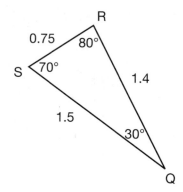

CPCTC

This isn't only a fun group of letters to say as quickly as you can, but it is a useful acronym to remember that **Corresponding Parts of Congruent Triangles are Congruent**. This fact is valuable when you are only given the *names* of two triangles without an illustration of them. Based on our discussion above on naming congruent triangles, you can identify congruent angles and sides by just using the triangles' names:

If $\triangle TOY \cong \triangle CAR$, then you know that the following congruencies are true:

$$\angle T \cong \angle C$$
$$\angle O \cong \angle A$$
$$\angle Y \cong \angle R$$

and

$$\overline{TO} \cong \overline{CA}$$
$$\overline{OY} \cong \overline{AR}$$
$$\overline{YT} \cong \overline{RC}$$

Proving Congruency in Triangles

Now that you understand the basics of congruency, it is time to learn how to know when two triangles are congruent. There are four different postulates, or tests, that prove congruency in triangles. These *postulates* investigate the **congruency** of corresponding angles and sides of triangles. If two triangles satisfy any of these four postulates, then they are congruent. Here are the four tests of congruency:

Test: 1 Side-Side-Side (SSS) Postulate: If three sides of one triangle are congruent to three sides of another triangle, then the two triangles are congruent.

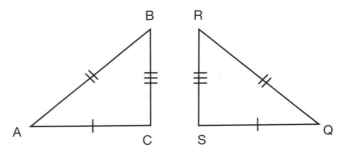

Test: 2 Side-Angle-Side (SAS) Postulate: If two sides and the included angle of one triangle are congruent to the corresponding sides and included angle of another triangle, then the triangles are congruent.

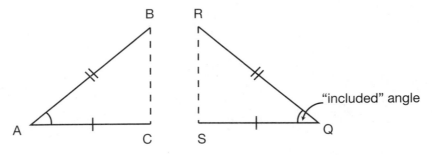

Test: 3 Angle-Side-Angle (ASA) Postulate: If two angles and the included side of one triangle are congruent to the corresponding two angles and the included side of another triangle, the triangles are congruent.

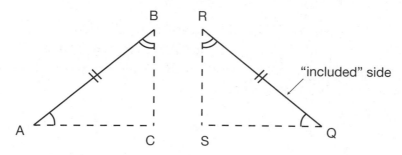

Test 4: Angle-Angle-Side (AAS) Postulate: If two angles and one non-included side of one triangle are congruent to the corresponding two angles and one non-included side of another triangle, then the triangles are congruent. (Note: in this case, the side does not need to be an included side between the two angles, like with the ASA postulate.)

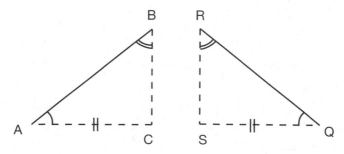

Warning! Although it is tempting to think that Angle-Angle-Angle would prove congruency, two triangles can have three congruent angles, but be completely different sizes. Therefore, Angle-Angle-Angle *cannot* be used to prove congruency, although it is a postulate used to prove **similarity**, which is one of the topics we will cover in the next chapter.

Set 27

Choose the best answer.

142. In ΔABC and ΔLMN, ∠A is congruent to ∠L, side BC is congruent to side MN, and side CA is congruent to side NL. Using the information above, which postulate proves that ΔABC and ΔLMN are congruent?
   a. SSS
   b. SAS
   c. ASA
   d. Congruency cannot be determined.

143. In ΔABC and ΔLMN, ∠C is congruent to ∠N, side BC is congruent to side MN, and side AC is congruent to side LN. Using the information above, which postulate proves that ΔABC and ΔLMN are congruent?
   a. SSS
   b. SAS
   c. ASA
   d. Congruency cannot be determined.

144. In ΔABC and ΔLMN, ∠A and ∠L are congruent, ∠B and ∠M are congruent and ∠C and ∠N are congruent. Using the information above, which postulate proves that ΔABC and ΔLMN are congruent?
   a. SSS
   b. SAS
   c. ASA
   d. Congruency cannot be determined.

Set 28

Use the figure below to answer questions 145 through 148.

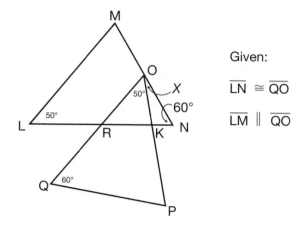

Given:

$\overline{LN} \cong \overline{QO}$

$\overline{LM} \parallel \overline{QO}$

145. Name each of the triangles in order of corresponding vertices.

146. Name the corresponding sides.

147. State the postulate that proves ΔLMN is congruent to ΔOPQ.

148. Find the measure of ∠X.

Set 29

Use the figure below to answer questions 149 through 152.

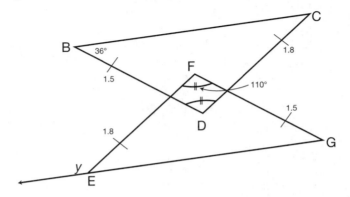

149. Name each of the triangles in order of corresponding vertices.

150. Name corresponding line segments.

151. State the postulate that proves ∆BCD is congruent to ∆EFG.

152. Find the measurement of ∠y.

Set 30

Use the figure below to answer questions 153 through 156.

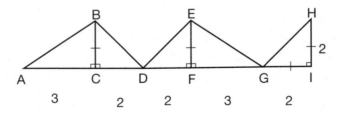

153. Name the three sets of congruent triangles in order of corresponding vertices.

154. Name corresponding line segments.

155. State the postulate that proves ∆ABC is congruent to ∆GEF.

156. How do you know that △ABD ≅ △GED?

Set 31

Use the figure below to answer questions 157 through 160. (Note: Figure is not drawn to scale.)

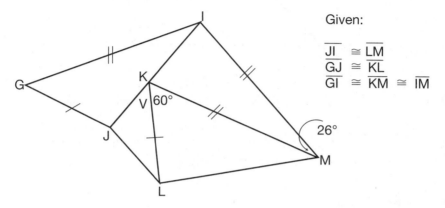

Given:

$\overline{JI} \cong \overline{LM}$
$\overline{GJ} \cong \overline{KL}$
$\overline{GI} \cong \overline{KM} \cong \overline{IM}$

157. Name a set of congruent triangles in order of corresponding vertices.

158. Name corresponding line segments.

159. State the postulate that proves △GIJ is congruent to △KML.

160. Find the measure of ∠V.

Set 32

Use the diagram below to answer questions 161 through 163.

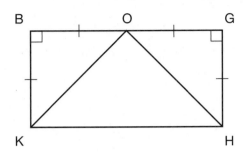

161. In the figure above, which triangles are congruent? What postulate proves it?

162. △HGO is what kind of special triangle?

163. ∠KOH measures _____ degrees.

Answers

Set 27

142. **d. Congruency cannot be determined.** In this example, you were given Angle-Side-Side which is a postulate that proves congruency.

143. **b. SAS.** The angle given is the included angle between the two sides given, and since all of the parts are corresponding, ΔABC and ΔLMN are congruent based on the Side-Angle-Side postulate.

144. **d. Congruency cannot be determined.** In this example, you were given Angle-Angle-Angle which is a postulate that proves *similarity*, but does not prove *congruency*. This fact was discussed in the last part of this chapter.

Set 28

145. **ΔLNM and ΔOQP**

Since ∠L = 50° and ∠QOP = 50°, then ∠L is corresponding to ∠O. Similarly since ∠N = 60°, and ∠Q = 60°, then ∠N is corresponding to ∠Q. Therefore, it is correct to say that ΔLNM corresponds to ΔOQP. Any order or direction will work when naming these triangles as long as angles L and O, N and Q, and M and P are corresponding.

146. $\overline{LM} \cong \overline{OP}$
$\overline{MN} \cong \overline{PQ}$
$\overline{NL} \cong \overline{QO}$

(Always coordinate corresponding endpoints.)

147. **Angle-Side-Angle postulate:**

∠N ≅ ∠Q
$\overline{LN} \cong \overline{QO}$
∠L ≅ ∠O

148. $x = 20.$ When a transversal crosses a pair of parallel lines, corresponding angles are congruent; so, ∠ORN measures 50°. ∠OKR measures 80°, and ∠OKR's supplement, ∠OKN, measures 100°. Finally, 180° – (100° + 60°) = 20°.

Set 29

149. ΔCDB and ΔEFG. (Remember to align corresponding vertices.)

150. $\overline{CD} \cong \overline{EF}$
$\overline{DB} \cong \overline{FG}$
$\overline{BC} \cong \overline{GE}$
(Always coordinate corresponding endpoints.)

151. Side-Angle-Side Postulate:

$\overline{BD} \cong \overline{GF}$
∠D ≅ ∠F
$\overline{CD} \cong \overline{EF}$

152. $m\angle y = 146°.$ Since ∠D = 110° and ∠B = 36°, it follows that ∠C = 34°. Since ∠C is corresponding to ∠FEG, then ∠FEG = 34°. ∠y is the supplementary angle to ∠FEG, and therefore $m\angle y = 180° – 34° = 146°.$

Set 30

153. There are three sets of congruent triangles in this question. ΔABC and ΔGEF make one set. ΔDBC, ΔDEF, and ΔGHI make the second set. ΔABD and ΔGED make the third set. (Remember to align corresponding vertices.)

154. Set one: $\overline{AB} \cong \overline{GE}, \overline{BC} \cong \overline{EF}, \overline{CA} \cong \overline{FG}$
Set two: $\overline{DB} \cong \overline{DE} \cong \overline{GH}$
$\overline{BC} \cong \overline{EF} \cong \overline{HI}$
$\overline{DC} \cong \overline{DF} \cong \overline{GI}$

155. Side-Angle-Side:
 Set one: $\overline{BC} \cong \overline{EF}$, $\angle BCA \cong \angle EFG$, $\overline{CA} \cong \overline{FG}$
 Set two: $\overline{BC} \cong \overline{EF} \cong \overline{HI}$
  $\angle BCD \cong \angle EFD \cong \angle I$
  $\overline{CD} \cong \overline{FD} \cong \overline{IG}$

156. $\overline{BC} \cong \overline{EF} \cong \overline{HI} = 2$ and also $\overline{DC} \cong \overline{DF} \cong \overline{GI} = 2$. Therefore
 ΔBCD and ΔEFD are right isosceles triangles with two base angles
 that each measure 45° and congruent hypotenuses. Since $\overline{BD} \cong$
 $\overline{ED}$, $\overline{DA} \cong \overline{DG}$, and $\angle BDA \cong \angle EDG$, $\Delta ABD \cong \Delta GED$ by the
 side-angle-side postulate.

Set 31

157. ΔKML and ΔGIJ. **(Remember to align corresponding
 vertices.)**

158. $\overline{KM} \cong \overline{GI}$
 $\overline{ML} \cong \overline{IJ}$
 $\overline{LK} \cong \overline{JG}$
 (Always coordinate corresponding endpoints.)

159. Side-Side-Side: $\overline{KM} \cong \overline{GI}$
  $\overline{ML} \cong \overline{IJ}$
  $\overline{LK} \cong \overline{JG}$

160. $m\angle V = 43°$. ΔIMK is an isosceles triangle. Its vertex angle
 measures 26°; its base angles measure 77° each. $180° - (m\angle IKM +
 m\angle MKL) = m\angle JKL$. $180° - (77° + 60°) = m\angle JKL$. $m\angle JKL = 43°$.

Set 32

161. ΔKBO and ΔHGO are congruent; Side-Angle-Side postulate.

162. Isosceles right triangle.

163. 90°. ΔBOK and ΔGOK are isosceles right triangles, $\angle BOK \cong$
 $\angle GOH = 45°$, so $\angle KOH$ must be 90°.

8

Ratio, Proportion, and Similarity

In the previous chapter it was mentioned that a 3 × 5 photo could be enlarged to fit on a billboard. In this case the images would be the same, but one would be much bigger than the other. In math, we call that type of likeness **similarity**. When two figures, such as triangles, have the same proportions between their respective angles and sides, but are different sizes, it is said that they are **similar triangles**. Remember the four postulates that can be used to prove that two triangles are congruent? Well there are three postulates that are used to demonstrate that triangles are similar. Two of these postulates are based on the ratios and proportions of the sides of the triangles, so before exploring them, we are going to discuss **ratios** and **proportions**.

Ratios and Proportions: The Basics

A **ratio** is a comparison of any two quantities. If I have 10 bikes and you have 20 cars, then the ratio of my bikes to your cars is 10 to 20. This ratio is simplified to 1 to 2 by dividing each side of the ratio by the greatest common factor (in this case, 10). Ratios are commonly written with a colon between the sets of objects being compared, but it is more useful in mathematical computations to write ratios as fractions. The ratio of bikes to cares is written bikes:cars = 10:20, which reduces to 1:2. This is said "one to two," meaning that for every one bike I have, you have two cars. This ratio can also be written as $\frac{10}{20} = \frac{1}{2}$.

A **proportion** is a statement that compares two equal ratios. Maybe the current ratio of my blue pens to my black pens is 7:2 or $\frac{7}{2}$. If I add four more black pens to my collection, a proportion can be used to determine how many blue pens I must add to maintain the same ratio of blue pens to black pens in my collection:

$$\frac{\text{blue pens}}{\text{black pens}} = \frac{7}{2} = \frac{p}{2+4} = \frac{p}{6}$$

Since the new denominator of six black pens is three times bigger than the original denominator of two black pens, I must multiply the seven blue pens in the numerator by three also, in order to maintain the same ratio. Therefore, I will need to have 21 blue pens, or 14 more blue pens than I had originally, if I want to maintain the same ration of blue to black pens. Notice that $\frac{21}{6}$ reduces to $\frac{7}{2}$, which was the original ratio.

Ratios and Proportions in Similar Triangles

If two triangles are **similar**, then the *triangles will be proportional*. What this means is that the ratios and proportions of their corresponding sides will be equal. This is useful because when dealing with similar triangles, proportions can be used to solve for unknown sides in the triangles. In the illustration below, $\triangle ABC$ is similar to $\triangle EDF$ and the symbol that is used to represent that is $\approx$. Therefore $\triangle ABC \approx \triangle EDF$. Based on the congruencies of the pairs of angles B and E, and C and F, it is clear that $\overline{BC}$ is proportional to $\overline{EF}$ and also that $\overline{BA}$ is proportional to $\overline{ED}$. With this information a proportion can be written as follows:

$$\frac{\overline{BA}}{\overline{BC}} = \frac{\overline{ED}}{\overline{EF}}$$
$$\frac{10}{6} = \frac{x}{12}$$

Since 12 is twice as large as six, it follows that x will have to be twice as large as 10, so $x = 20$. If $\overline{DF}$ was 18 units long, it would follow that $\overline{AC}$ would be nine since $\triangle ABC$ is half as big as $\triangle EDF$.

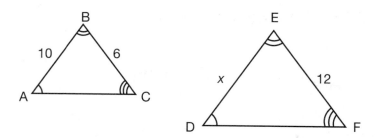

CAUTION! When working with ratios and proportions, it doesn't matter which number goes in the numerator and which goes in the denominator in the first ratio. However, when you are writing out the second fraction, you must be sure to line up the corresponding parts correctly. For example, looking at the triangles above, the following proportion is **incorrect** since the corresponding sides are not lined up correctly:

$$\frac{\overline{BA}}{\overline{BC}} \neq \frac{\overline{EF}}{\overline{ED}}$$

$$\frac{10}{6} \neq \frac{12}{x}$$

The illustration below shows all of the proportional relationships between the similar triangles, $\triangle AEB$ and $\triangle CDB$. Notice that since the triangles are similar, with $\overline{CB} \approx \overline{AB}$ and $\overline{AB}$ being twice as big as $\overline{CB}$, it can be concluded that the relationship between the other pairs of corresponding sides will also be 1:2.

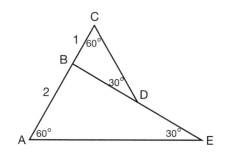

Corresponding Angles of Similar Triangles
Are Congruent (CASTC)

$\angle A \cong \angle C$

$\angle ABE \cong \angle CBD$

$\angle CDB \cong \angle AEB$

Corresponding Sides of Similar Triangles
Are Proportional (CPSTP)

$2 \times \overline{BC} = 1 \times \overline{AB}$

$2 \times \overline{BD} = 1 \times \overline{BE}$

$2 \times \overline{CD} = 1 \times \overline{AE}$

Different sizes
Same shape
Different measurements, but in proportion

Proving that Triangles are Similar

Not only can proportions be used to solve for missing sides when it is *known* that two triangles are similar, but conversely, proportions can be used to determine if two triangles are **similar**. There are three different *postulates*, or tests, that prove **similarity** in triangles. These postulates investigate the measurements of corresponding angles as well as the proportions of corresponding sides in triangles. If two triangles satisfy any of these three postulates, then they are similar. Here are the three tests of similarity:

Test 1: Angle-Angle (AA) Postulate: If two angles of one triangle are congruent to two angles of another triangle, then the triangles are similar. This is true because two angles really prove that all three angles are equal, since their sum needs to be 180°.

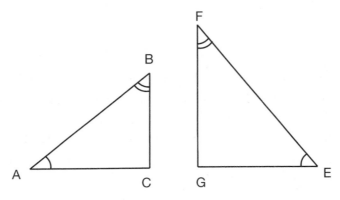

Test 2: Side-Side-Side (SSS) Postulate: If the lengths of the corresponding sides of two triangles are proportional, then the triangles are similar.

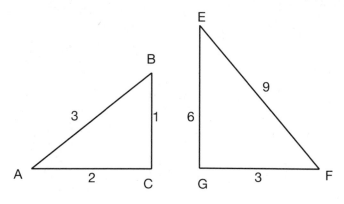

See Ratios and Proportions

$$\overline{AB} : \overline{EF} = 3:9$$

$$\overline{BC} : \overline{FG} = 1:3$$

$$\overline{CA} : \overline{GE} = 2:6$$

$$3:9 = 2:6 = 1:3$$

Reduce each ratio,

$$1:3 = 1:3 = 1:3$$

Test 3: Side-Angle-Side (SAS) Postulate: If the lengths of two pairs of corresponding sides of two triangles are proportional and the corresponding included angles are congruent, then the triangles are similar.

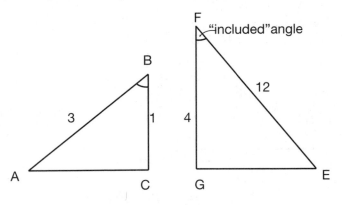

See Ratios and Proportions

$$\overline{AB} : \overline{EF} = 3:12$$

$$\overline{BC} : \overline{FG} = 1:4$$

$$3:12 = 1:4$$

Reduce each ratio,

$$1:4 = 1:4$$

Set 33

Choose the best answer.

164. If ΔDFG and ΔJKL are both right and isosceles, which postulate proves they are similar?
a. Angle-Angle
b. Side-Side-Side
c. Side-Angle-Side
d. Angle-Side-Angle

165. In ΔABC, side AB measures 16 inches. In similar ΔEFG, corresponding side EF measures 24 inches. State the ratio of side AB to side EF.

   a. 2:4

   b. 2:3

   c. 2:1

   d. 8:4

166. Use the figure below to find a proportion to solve for *x*.

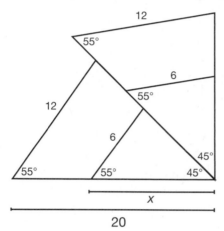

   a. $\frac{12}{6} = \frac{20}{(20 - x)}$

   b. $\frac{12}{20} = \frac{x}{6}$

   c. $\frac{20}{12} = \frac{6}{x}$

   d. $\frac{12}{6} = \frac{20}{x}$

167. In similar ΔUBE and ΔADF, $\overline{UB}$ measures 10 inches while corresponding $\overline{AD}$ measures 2 inches. If $\overline{BE}$ measures 30 inches, then corresponding $\overline{DF}$ measures

   a. 150 inches.

   b. 60 inches.

   c. 12 inches.

   d. 6 inches.

Set 34

Use the figure below to answer questions 168 through 171.

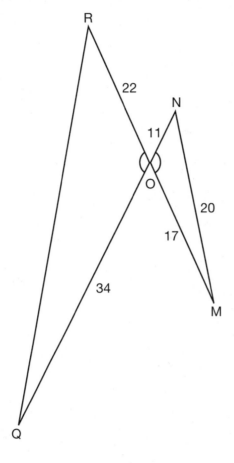

168. Name each of the triangles in order of their corresponding vertices.

169. Name corresponding line segments.

170. State the postulate that proves similarity.

171. Find $\overline{RQ}$.

Set 35

Use the figure below to answer questions 172 through 175.

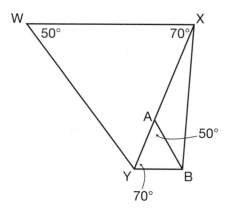

172. Name the similar triangles in order of corresponding vertices.

173. $\overline{YB} = 3$, $\overline{YW} = 16$. amd $\overline{XY} = 12$. Find $\overline{AB}$.

174. State the postulate that proves similarity.

175. Prove that $\overline{WX}$ and $\overline{YB}$ are parallel.

Set 36

Use the figure below to answer questions 176 through 179.

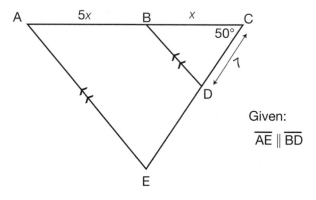

Given:
$\overline{AE} \parallel \overline{BD}$

176. Name a pair of similar triangles in order of corresponding vertices.

177. Name corresponding line segments.

178. State the postulate that proves similarity.

179. Find $\overline{EC}$.

Set 37

Fill in the blanks with a letter from a corresponding figure in the box below.

Triangle A

Triangle B

Triangle C

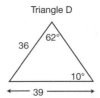

Triangle D

Triangle E

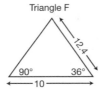

Triangle F

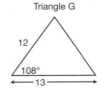

Triangle G

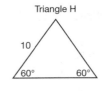

Triangle H

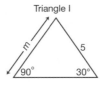

Triangle I

Triangle J

Triangle K

Triangle L

180. Choice _____ is congruent to ΔA.

181. Choice _____ is similar to ΔA.

182. Choice _____ is congruent to ΔB.

183. Choice _____ is similar to ΔB.

184. Choice _____ is congruent to ΔE.

185. Choice _____ is similar to ΔE.

186. Choice _____ is congruent to ΔD.

187. Choice _____ is similar to ΔD.

188. Find the length of side m in ΔI.

189. Find the length of side r in ΔE.

Answers

Set 33

164. **a.** The angles of a right isosceles triangle always measure 45° – 45° – 90°. Since at least two corresponding angles are congruent, right isosceles triangles are similar.

165. **b.** A ratio is a comparison. If one side of a triangle measures 16 inches, and a corresponding side in another triangle measures 24 inches, then the ratio is 16:24. This ratio can be simplified by dividing each side of the ratio by the common factor 8. The comparison now reads, 2:3 or *2 to 3*. Choices **a**, **c**, and **d** simplify into the same incorrect ratio of 2:1 or 1:2.

166. **d.** When writing a proportion, corresponding parts must parallel each other. The proportions in choices **b** and **c** are misaligned. Choice **a** looks for the line segment 20 – *x*, not *x*.

167. **d.** First, state the ratio between similar triangles; that ratio is 10:2 or 5:1. The ratio means that a line segment in the larger triangle is always 5 times more than the corresponding line segment in a similar triangle. If the line segment measures 30 inches, it is 5 times more than the corresponding line segment. Create the equation: 30 = 5*x*. *x* = 6.

Set 34

168. ΔOQR and ΔOMN. (Remember to align corresponding vertices.)

169. **Corresponding line segments are $\overline{OQ}$ and $\overline{OM}$; $\overline{QR}$ and $\overline{MN}$; $\overline{RO}$ and $\overline{NO}$.** Always coordinate corresponding endpoints.

170. **Side-Angle-Side.** The sides of similar triangles are not congruent; they are proportional. If the ratio between corresponding line-segments, $\overline{RO}$ and $\overline{NO}$ is 22:11, or 2:1, and the ratio between corresponding line segments $\overline{QO}$ and $\overline{MO}$ is also 2:1, they are proportional.

171. $x = 40.$ From the last question, you know the ratio between similar triangles OQR and OMN is 2:1. That ratio means that a line segment in the smaller triangle is half the size of the corresponding line segment in the larger triangle. If that line segment measures 20 inches, it is half the size of the corresponding line segment. Create the equation: $20 = \frac{1}{2}x.$ $x = 40.$

Set 35

172. **ΔWXY and ΔAYB.** (Remember to align corresponding vertices.)

173. $\overline{AB} = 4.$ Since $\overline{WY} \approx \overline{AB}$, set up the proportion:

$\frac{\overline{WY}}{\overline{XY}} = \frac{\overline{AB}}{\overline{YB}}, \frac{16}{12} = \frac{\overline{AB}}{3}$

Since $\overline{YB}$ is $\frac{1}{4}$ the length of $\overline{XY}$, $\overline{AB}$ will be $\frac{1}{4}$ the length of WY. $\frac{1}{4}$ of 16 is 4, so $\overline{AB} = 4.$

174. **Angle-Angle postulate.** Since there are no side measurements to compare, only an all-angular postulate can prove triangle similarity.

175. $\overline{XY}$ **acts like a transversal across** $\overline{WX}$ **and** $\overline{BY}$. When alternate interior angles are congruent, then lines are parallel. In this case, ∠WXY and ∠BYA are congruent alternate interior angles. $\overline{WX}$ and $\overline{BY}$ are parallel.

Set 36

176. **ΔAEC and ΔBDC.** (Remember to align corresponding vertices.)

177. **Corresponding line segments are** $\overline{AE}$ **and** $\overline{BD}$; $\overline{EC}$ **and** $\overline{DC}$; $\overline{CA}$ **and** $\overline{CB}$. Always coordinate corresponding endpoints.

178. **Angle-Angle postulate.** Though it is easy to overlook, vertex C applies to both triangles.

179. $\overline{EC}$ = 42. First redraw both of the similar triangles. Since $\overline{AC}$ = $\overline{AB}$ + $\overline{BC}$, $\overline{AC}$ will be:

$$\frac{\overline{BC}}{\overline{CD}} = \frac{\overline{AC}}{\overline{EC}}$$

$$\frac{6x}{7} = \frac{6x}{\overline{EC}}$$

Since $\overline{AC}$ is six times largere than $\overline{BC}$, $\overline{EC}$ will be six times larger than $\overline{CD}$. Therefore, $\overline{EC}$ = 42.

Set 37

180. Δ**C.** Because the two angles given in ΔA are 30° and 60°, the third angle in ΔA is 90°. Like ΔA, choices **c** and **i** also have angles that measure 30°, 60°, and 90°. According to the Angle-Angle postulate, at least two congruent angles prove similarity. To be congruent, an included side must also be congruent. ΔA and ΔC have congruent hypotenuses. They are congruent.

181. Δ**I.** In the previous answer, ΔC was determined to be congruent to ΔA because of congruent sides. In ΔI, the triangle's hypotenuse measures 5; it has the same shape as ΔA but is smaller; consequently, they are not congruent triangles; they are only similar triangles.

182. Δ**K.** ΔB is an equilateral triangle. ΔH and ΔK are also equilateral triangles (an isosceles triangle whose vertex measures 60° must also have base angles that measure 60°). However, only ΔK and ΔB are congruent because of congruent sides.

183. Δ**H.** ΔH has the same equilateral shape as ΔB, but they are different sizes. They are not congruent; they are only similar.

184. Δ**J.** The three angles in ΔE measure 36°, 54°, and 90°. ΔF and ΔJ also have angles that measure 36°, 54°, and 90°. According to the Angle-Angle postulate, at least two congruent angles prove similarity. To be congruent, an included side must also be congruent. The line segments between the 36° and 90° angles in ΔJ and ΔE are congruent.

185. Δ**F.** ΔF has the same right scalene shape as ΔE, but they are not congruent; they are only similar.

186. Δ**L.** The three angles in ΔD respectively measure 62°, 10°, and 108°. ΔL has a set of corresponding and congruent angles, which proves similarity; but ΔL also has an included congruent side, which proves congruency.

187. Δ**G.** ΔG has only one given angle; the Side-Angle-Side postulate proves it is similar to ΔD. The sides on either side of the 108° angle are proportional and the included angle is obviously congruent.

188. *m* **= 2.5.** Since ΔI ≈ ΔA, set up the following proportion to solve for m, which pairs the two sides opposite the 90° angles and the two sides that are opposite the 30° angles:

$$\frac{20}{10} = \frac{5}{m} =$$

M will be $\frac{1}{4}$ of 10, so $(\frac{1}{4}) \times 10 = \frac{10}{4} = 2.5$.

189. *r* **= 6.2.** Since ΔF ≈ ΔE, set up the following proportion which pairs up the two sides that are opposite the 90° angles and the two sides that are opposite the 54° angles:

$$\frac{10}{12.4} = \frac{5}{r}$$

Since ΔE is half as big as ΔF, *r* will be equal to 6.2.

Triangles and the Pythagorean Theorem

In Chapters 7 and 8, you found the unknown sides of a triangle using proportions and the known sides of similar and congruent triangles. To find an unknown side of a single right triangle, you will need to use the Pythagorean theorem.

The Pythagorean theorem is a special equation that defines the relationship between the sides of **right triangles**. In order to use it successfully, you must first learn the standard way to label the sides of right triangles. In a right triangle, the two sides that form the right angle are referred to as the **legs**. It is standard for the legs of a right triangle to be called *a* and *b*. The longest side (which is always opposite the right angle), is called the **hypotenuse** and it is standard for the hypotenuse to be called *c*. The correct standard labeling of a right triangle is shown in the following illustration:

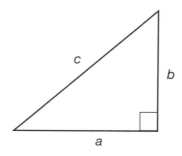

Pythagorean Theorem

The **Pythagorean theorem** is used when you know the lengths of two sides of a right triangle and want to find the length of the missing third side. This can be done when the two given lengths are the legs, or when you know the length of the hypotenuse and one of the legs. What is most important to remember, is that in your equation, the hypotenuse is **always** the side that begins all alone. Here's the formula:

(length of leg #1)^2 + (length of leg #2)^2 = (length of hypotenuse)^2

Since it is standard to label the sides of the triangle as **a**, **b**, and **c** as in the previous figure, the Pythagorean theorem is more commonly known as:

$$a^2 + b^2 = c^2$$

A note on exponents! If you are not familiar with the small "2" that is above and to the right of each letter, this is an exponent. "**a^2**" does not mean a times two, but it means **a** times itself _two times_. Therefore, "**x^3**" would mean _x_ times itself _three_ times and **not _x_ times three**. To undo an exponent, you need to take the square root of a number, which means discovering what number multiplied by itself two times gives that answer. The symbol for the square root looks like $\sqrt{}$. Therefore, $x^2 = \sqrt{64}$ means what number, multiplied by itself two times, yields 64. (The answer is 8, since $\sqrt{64} = 8$).

Look carefully over the following three examples to see how the Pythagorean theorem is used with right triangles:

Example 1: Find hypotenuse $\overline{QR}$.

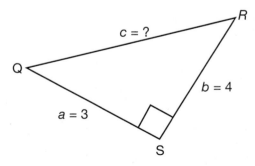

$$a^2 + b^2 = c^2$$
$$3^2 + 4^2 = c^2$$
$$9 + 16 = c^2$$
$$25 = c^2$$

Take the square root of each side:

$$\sqrt{25} = \sqrt{c^2}$$
$$5 = c$$

Example 2: Find $\overline{KL}$.

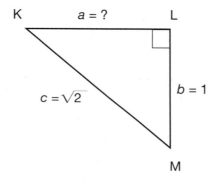

$$a^2 + b^2 = c^2$$
$$a^2 + 1^2 = (\sqrt{2})^2$$
$$a^2 + 1 = 2$$
$$a^2 = 1$$

Take the square root of each side:

$$\sqrt{a^2} = \sqrt{1}$$

Example 3: Find $\overline{CD}$.

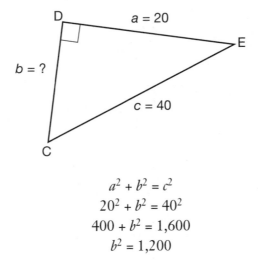

$$a^2 + b^2 = c^2$$
$$20^2 + b^2 = 40^2$$
$$400 + b^2 = 1,600$$
$$b^2 = 1,200$$

Take the square root of each side:

$$\sqrt{b^2} = \sqrt{1,200}$$
$$b = 20\sqrt{3}$$

Additional Application of Pythagorean Theorems

The Pythagorean theorem can only find a side of a right triangle. However, if all the sides of any given triangle are known, but none of the angles are known, the Pythagorean theorem can tell you whether that triangle is obtuse or acute.

If the square of the longest side is *larger* than the sum of the squares of two smaller sides, then the longest side is creating an **obtuse** angle between the two smaller sides and the triangle is obtuse. Conversely, if the square of the longest side is *smaller* than the sum of the squares of two smaller sides, then the longest side creates an **acute** angle between the two smaller sides and the triangle is acute.

Caution! Before using the Pythagorean theorem to determine if triangles are acute or obtuse in the following two examples, remember this important rule about the sides of triangles: the sum of any two sides of a triangle must be longer than the third side, or else the triangle cannot exist.

Is △GHI obtuse or acute?

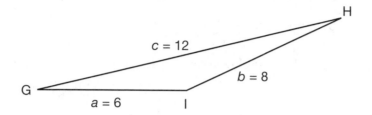

$a^2 + b^2$	c^2
$6^2 + 8^2$	12^2
36 + 64	144

100   <   144

Therefore, △GHI is obtuse.

Is △JKL obtuse or acute?

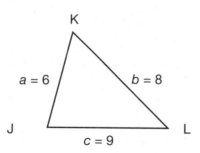

$a^2 + b^2$	c^2
$6^2 + 8^2$	9^2
36 + 64	81

100   >   81

Therefore, △JKL is acute.

Set 38

Choose the best answer.

190. If the sides of a triangle measure 3, 4, and 5, then the triangle is
    a. acute.
    b. right.
    c. obtuse.
    d. It cannot be determined.

191. If the sides of a triangle measure 12, 13, and 16, then the triangle is
    a. acute.
    b. right.
    c. obtuse.
    d. It cannot be determined.

192. If the sides of a triangle measure 15, 17, and 22, then the triangle is
    a. acute.
    b. right.
    c. obtuse.
    d. It cannot be determined.

193. If the sides of a triangle measure 8, 10, and 18, then the triangle is
    a. acute.
    b. right.
    c. obtuse.
    d. It cannot be determined.

194. If the sides of a triangle measure 12, 12, and 15, then the triangle is
    a. acute.
    b. right.
    c. obtuse.
    d. It cannot be determined.

195. If two sides of a triangle measure 4 and 14, and an angle measures 34°, then the triangle is
    a. acute.
    b. right.
    c. obtuse.
    d. It cannot be determined.

196. If the sides of a triangle measure 12, 16, and 20, then the triangle is
  a. acute.
  b. right.
  c. obtuse.
  d. It cannot be determined.

Set 39

Choose the best answer.

197. Eva and Carr meet at a corner. Eva turns 90° left and walks 5 paces; Carr continues straight and walks 6 paces. If a line segment connected them, it would measure
  a. $\sqrt{22}$ paces.
  b. $\sqrt{25}$ paces.
  c. $\sqrt{36}$ paces.
  d. $\sqrt{61}$ paces.

198. The legs of a square table measure 3 feet long and the top edges measure 4 feet long. If the legs are connected to the table at a right angle, then what is the diagonal distance from the bottom of a leg to the adjacent corner of the tabletop?
  a. 5 feet
  b. 7 feet
  c. 14 feet
  d. 25 feet

199. Dorothy is standing 300 meters directly below a plane. She sees another plane flying straight behind the first. It is 500 meters away from her, and she has not moved. How far apart are the planes from each other?
  a. 40 meters
  b. 400 meters
  c. 4,000 meters
  d. 40,000 meters

200. A surveyor is using a piece of equipment to measure a rectangular plot of property. Standing on one of the corners she sights the corner property boundary directly north of her as being 10 meters away. Then she turns 90-degrees to her right and sights the east-most corner as being 15 meters away. How far is the surveyor from the northeast corner that is diagonally opposite from where she is standing?

    a. 35 feet.

    b. 50 feet.

    c. $\sqrt{225}$ feet.

    d. $\sqrt{325}$ feet.

Set 40

Use the figure below to answer questions 201 through 203.

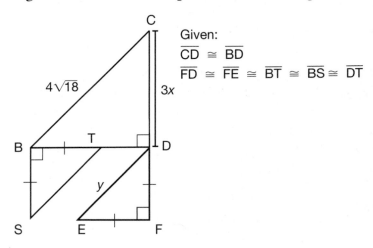

Given:
$\overline{CD} \cong \overline{BD}$
$\overline{FD} \cong \overline{FE} \cong \overline{BT} \cong \overline{BS} \cong \overline{DT}$

201. Which triangles in the figure above are congruent and/or similar?

202. Find the value of x.

203. Find the value of y.

Set 41

Use the figure below to answer questions 204 through 206.

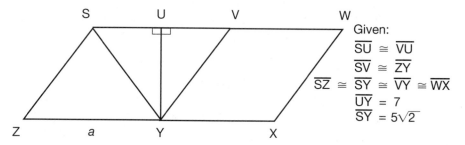

Given:

$\overline{SU} \cong \overline{VU}$

$\overline{SV} \cong \overline{ZY}$

$\overline{SZ} \cong \overline{SY} \cong \overline{VY} \cong \overline{WX}$

$\overline{UY} = 7$

$\overline{SY} = 5\sqrt{2}$

204. Find $\overline{UV}$

205. Find the value of a.

206. Is $\triangle ZSY$ acute or obtuse?

Set 42

Use the figure below to answer questions 207 through 209.

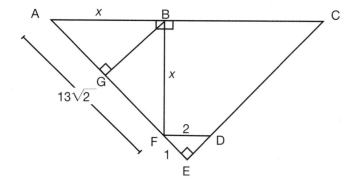

Given:

$\overline{AE} \cong \overline{CE}$

$\overline{FE} \cong \overline{ED}$

$\overline{AG} \cong \overline{BG}$

207. Find the value of x.

208. Find $\overline{AG}$.

209. Find $\overline{AC}$.

Set 43

Use the figure below to answer questions 210 through 215.

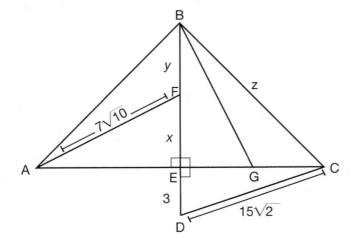

Given:

$\overline{AE} \cong \overline{CE}$

$\overline{AF} \cong \overline{BG}$

$\overline{AB} \cong \overline{CB}$

$\overline{BF} \cong \overline{CG}$

$\overline{FE} \cong \overline{GE}$

$\overline{EC} = w$

210. Which triangles in the figure above are congruent and/or similar?

211. Find the value of w.

212. Find the value of x.

213. Find the value of y.

214. Find the value of z.

215. Is $\triangle BGC$ acute or obtuse?

Answers

Set 38

190. **b.** This is a popular triangle, so know it well. A 3-4-5 triangle is a right triangle. Apply the Pythagorean theorem: $a^2 + b^2 = c^2$. $3^2 + 4^2 = 5^2$. $9 + 16 = 25$. $25 = 25$.

191. **a.** Plug the given measurements into the Pythagorean theorem: $12^2 + 13^2 = 16^2$. $144 + 169 = 256$. $313 > 256$. Acute.

192. **a.** Plug the given measurements into the Pythagorean theorem (the largest side is always c in the theorem): $15^2 + 17^2 = 22^2$. $225 + 289 = 484$. $514 > 484$. When the sum of the smaller sides squared is greater than the square of the largest side, then the triangle is acute.

193. **d.** In order for a triangle to exist, the sum of the length of any two sides needs to be longer than the length of third side. The given sides, 8, 10, and 18, do not make a triangle since the sum of the smaller two sides is only 18, which is equal to, but not greater than, the third side.

194. **a.** Plug the given measurements into the Pythagorean theorem: $12^2 + 12^2 = 15^2$. $144 + 144 = 225$. $288 > 225$. Acute.

195. **d.** The Pythagorean theorem does not include any angles. Without a third side or a definite right angle, this triangle cannot be determined.

196. **b.** This is also a 3–4–5 triangle. Simplify the measurement of each side by dividing 12, 16, and 20 by 4: $\frac{12}{4} = 3$. $\frac{16}{4} = 4$. $\frac{20}{4} = 5$.

197. **d.** The corner forms the right angle of this triangle; Eva and Carr walk the distance of each leg, and the question wants to know the hypotenuse. Plug the known measurements into the Pythagorean theorem: $5^2 + 6^2 = c^2$. $25 + 36 = c^2$. $61 = c^2$. $\sqrt{61} = c$.

198. **a.** The connection between the leg and the tabletop forms the right angle of this triangle. The length of the leg and the length of the top are the legs of the triangle, and the question wants to know the distance of the hypotenuse. Plug the known measurements into the Pythagorean theorem: $3^2 + 4^2 = c^2$. $9 + 16 = c^2$. $25 = c^2$. $5 = c$. If you chose answer **d**, you forgot to take the square root of 25. If you chose answer **b**, you added the legs together without squaring them first.

199. **b.** The first plane is actually this triangle's right vertex. The distance between Dorothy and the second plane is the hypotenuse. Plug the known measurements into the Pythagorean theorem: $300^2 + b^2 = 500^2$. $90,000 + b^2 = 250,000$. $b^2 = 160,000$. $b = 400$. Notice that if you divided each side by 100, this is another 3-4-5 triangle.

200. **d.** The bases of Timmy's walls form the legs of this right triangle. The hypotenuse is unknown. Plug the known measurements into the Pythagorean theorem: $10^2 + 15^2 = c^2$. $100 + 225 = c^2$. $325 = c^2$. $\sqrt{325} = c$.

Set 40

201. $\triangle$SBT and $\triangle$EFD are congruent to each other (Side-Angle-Side theorem) and similar to $\triangle$BDC (Angle-Angle theorem).

202. $x = 4$. Because $\triangle$BCD is an isosceles right triangle, $\overline{BD}$ is congruent to $\overline{CD}$. Plug $3x$, $3x$, and $4\sqrt{18}$ into the Pythagorean theorem: $(3x)^2 + (3x)^2 = (4\sqrt{18})^2$. $9x^2 + 9x^2 = 288$. $18x^2 = 288$. $x^2 = 16$. $x = 4$.

203. $y = 6\sqrt{2}$. In the question above, you found $x = 4$. Therefore, CD = 12. Since BT = DT, they both equal 6. Since BT = FD = FE, FD = FE = 6. Plug 6, 6, and y into the Pythagorean theorem.

204. Remember $\overline{SU} \cong \overline{UV}$, so to find the measurement of $\overline{SU}$, plug the given measurements of $\triangle$SUY into the Pythagorean theorem. $7^2 + b^2 = (5\sqrt{2})^2$. $49 + b^2 = 50$. $b^2 = 1$. $b = \sqrt{1} = 1$.

Set 41

205. $a = 2$. $\overline{SU} + \overline{UV} = \overline{ZY}$. $\overline{SU} = \overline{UV}$. Since $\overline{SU} = 1$ as demonstrated in question 204, $\overline{ZY} = 1 + 1 = 2$.

206. **Acute.** $\triangle ZSY$ is an isosceles triangle. Two of its sides measure $5\sqrt{2}$. The third side measures 2. Plug the given measures into the Pythagorean theorem. $2^2 + (5\sqrt{2})^2 = (5\sqrt{2})^2$. Thus, $4 + 50 = 50$; $54 > 50$. Therefore, $\triangle ZSY$ is acute.

Set 42

207. $x = 13$. Even though you don't know the measurement of x in $\triangle ABF$, you do know that two sides measure x. Plug the measurements of $\triangle ABF$ into the Pythagorean theorem. $x^2 + x^2 = (13\sqrt{2})^2$. $2x^2 = 338$. $x^2 = 169$. $x = 13$.

208. $AG = \sqrt{\frac{169}{2}}$. Since $\overline{AG} \cong \overline{BG}$ and $\angle AGB = 90°$, plug "a" into the Pythagorean theorem for both legs: $a^2 + a^2 = 13^2$, $2a^2 = 169$, $a = \sqrt{\frac{169}{2}}$.

209. $26\sqrt{2} + 2$. The ratio between corresponding line segments $\overline{AE}$ and $\overline{FE}$ is $13\sqrt{2} + 1{:}1$. Since $\overline{FD} = 2$, AC is twice the size of AE.

Set 43

210. $\triangle AFE$ and $\triangle BGE$ are congruent (Side-Side-Side postulate). $\triangle ABF$ and $\triangle BCG$ are congruent (Side-Side-Side postulate).

211. $w = 21$. Plug the measurements of $\triangle ECD$ into the Pythagorean theorem: $3^2 + w^2 = (15\sqrt{2})^2$. $9 + w^2 = 450$. $w^2 = 441$. $w = 21$.

212. $x = 7$. Corresponding parts of congruent triangles are congruent (CPCTC). If $\overline{EC}$ is 21, then $\overline{EA}$ is also 21. Plug the measurements of $\triangle AFE$ into the Pythagorean theorem: $21^2 + x^2 = (7\sqrt{10})^2$. $441 + x^2 = 490$. $x^2 = 49$. $x = 7$.

213. $y = 14$. Because of CPCTC, $\overline{AE}$ is also congruent to $\overline{BE}$. If $\overline{BE}$ is 21 and $\overline{FE}$ is 7, subtract 7 from 21 to find $\overline{BF}$. $21 - 7 = 14$.

214. $z = 21\sqrt{2}$. Plug the measurements of $\triangle BEC$ into the Pythagorean theorem: $21^2 + 21^2 = z^2$. $441 + 441 = z^2$. $882 = z^2$. $21\sqrt{2} = z$.

215. **Obtuse.** You could just guess that $m\angle BGC > 90°$. However, the question wants you to use the Pythagorean theorem to show $(7\sqrt{10})^2 + 14^2 < (21\sqrt{2})^2$.

10

Properties of Polygons

Polygons are closed geometric figures that have three or more straight line segments as sides. Two sides of a polygon are **adjacent sides** if they are next to each other and share an angle. Similarly, **adjacent angles** are angles that are next to one another, separated only by one side of the polygon. Polygons have **interior angles** at each of their vertices. The more sides a polygon has, the larger the sum of its interior angles becomes. For example, the sum of the interior angles of a triangle is 180° and the sum of a rectangle's interior angles is 360°. Polygons have **diagonals**, which are line segments that connect any two *non-adjacent* angles.

Naming Polygons

Polygons are named by listing all of their vertices in clockwise or counter-clockwise order. Polygons are most basically classified by the number of sides they have. The classification can become more specific based on information about a polygon's angles or sides. For example, a **regular polygon** is equilateral (has all sides of equal length) as well as equiangular (all interior angles are equal in measure). The following figure is regular octagon ABCDEFGH ("octa" means eight, so an octagon is an eight-sided polygon).

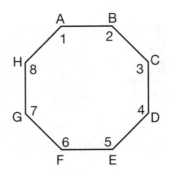

sides: $\overline{AB} = \overline{BC} = \overline{CD} = \overline{DE} = \overline{EF} = \overline{FG} = \overline{GH} = \overline{HA}$
interior $\angle$s: $\angle 1 \cong \angle 2 \cong \angle 3 \cong \angle 4 \cong \angle 5 \cong \angle 6 \cong \angle 7 \cong \angle 8$

Vertices of a **convex polygon** all point outwards. All of the interior angles of convex angles are less than 180°. Pentagon ABCDE below is convex ("penta" means five, so a pentagon is a five-sided polygon).

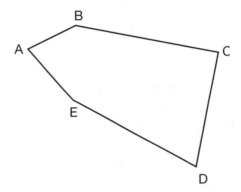

$m \angle A, \angle B, \angle C, \angle D, \angle E, < 180$

If any of the vertices of a polygon point inward or if the measure of any vertex exceeds 180°, the polygon is a **concave polygon**. In heptagon ABCDEFG below, $\angle D$ is greater than 180°, so it is a concave polygon.

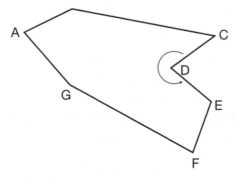

$m \angle D > 180$, therefore polygon ABCDEFG is concave.

Finally, let's discus how polygons are named by the number of sides they have. By now you are familiar with three-sided polygons or **triangles**. A regular triangle is an *equilateral triangle*. A four-sided polygon is called a **quadrilateral**. A regular quadrilateral is a *square*. Polygons with five sides or more take special prefixes. Study these names below:

Five-sided	PENTAgon
Six-sided	HEXAgon
Seven-sided	HEPTAgon
Eight-sided	OCTAgon
Nine-sided	NONAgon
Ten-sided	DECAgon
Twelve-sided	DODECAgon

SET 44

State whether the object is or is not a polygon and why. (Envision each of these objects as simply as possible, otherwise there will always be exceptions.) If the object is a polygon, name the polygon as specifically as possible.

216. a picture frame for a 5 × 7 photo

217. Manhattan's grid of city blocks

218. branches of a tree

219. the block letter "M" carved into the tree

220. half of a peanut butter and jelly sandwich that was cut diagonally into equal parts

221. a balloon

222. a stop sign

223. lace

Set 45

Use the diagram below to answer questions 224 through 226.

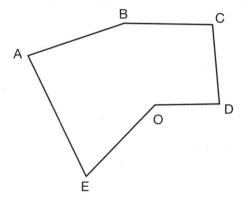

224. Name the polygon. Is it convex or concave?

225. How many diagonals can be drawn from vertex O?

226. Based on the number of triangles that are created from the diagonals drawn from vertex O, what is the sum of the interior angles of this 6-sided polygon?

Set 46

Use the diagram below to answer questions 227 through 229.

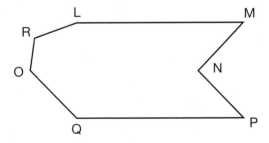

227. Name the polygon. Is it convex or concave?

228. How many diagonals can be drawn from vertex O?

229. Based on the number of triangles created from the diagonals drawn from vertex O, what is the sum of the interior angles of this 7-sided polygon?

Set 47

Use the diagram below to answer questions 230 through 232.

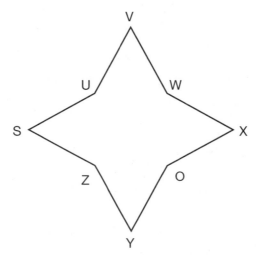

230. Name the polygon. Is it convex or concave?

231. How many diagonals can be drawn from vertex O?

232. Based on the number of triangles created from the diagonals drawn from vertex O, what is the sum of the interior angles of this 8-sided polygon?

Set 48

Use the diagram below to answer questions 233 through 235.

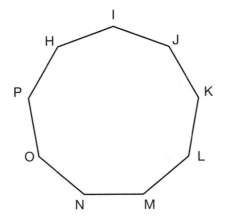

Given:
$\overline{HI} \cong \overline{IJ} \cong \overline{JK} \cong$
$\overline{KL} \cong \overline{LM} \cong \overline{MN} \cong$
$\overline{NO} \cong \overline{OP} \cong \overline{PH}$

233. Name the polygon. Is it convex or concave?

234. How many diagonals can be drawn from vertex O?

235. Based on the number of triangles created from the diagonals drawn from vertex O, what is the sum of the interior angles of this 9-sided polygon?

Set 49

Use your knowledge of polygons to fill in the blanks.

236. In polygon CDEFG, $\overline{CD}$ and $\overline{DE}$ are _____.

237. In polygon CDEFG, $\overline{CE}$, $\overline{DF}$ and $\overline{EG}$ are _____.

238. In polygon CDEFG, ∠EFG is also named _____.

239. In polygon CDEFG, ∠DEF and ∠EFG are _____.

Set 50

Use diagonals to draw the triangles below.

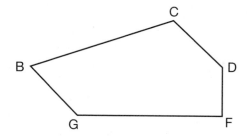

240. How many non-overlapping triangles can be drawn in the accompanying polygon at one time?

241. Determine the sum of the polygon's interior angles by using the number of triangles; verify your answer by using the formula $s = 180°(n - 2)$, where s is the sum of the interior angles and n is the number of sides the polygon has.

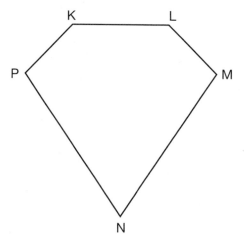

242. How many non-overlapping triangles can be drawn in the accompanying polygon at one time?

243. Determine the sum of the polygon's interior angles using the number of triangles; then apply the formula $s = 180°(n - 2)$ to verify your answer.

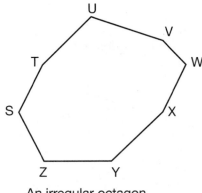

An irregular octagon

244. How many non-overlapping triangles can be drawn in the
accompanying polygon at one time?

245. Determine the sum of the polygon's interior angles using the
number of triangles; then apply the formula $s = 180°(n - 2)$ to
verify your answer.

Answers

Set 44

216. **Rectangle.** A frame is a quadrilateral with four 90-degree angles.

217. **Not a polygon.** A grid is not a polygon because its lines intersect at points that are not endpoints.

218. **Not a polygon.** Branches are open, and they "branch" out at points that are also not endpoints.

219. **Concave Dodecagon.** A block letter is a closed multi-sided figure; each of its line segments begin and end at an endpoint. A block letter "M" will not have any curves and it will have 12 sides and three interior angles that exceed 180°.

220. **Triangle** If the bread was square to begin, then it would be a right isosceles triangle. If it was a rectangular slice of bread then it would be a right scalene triangle.

221. **Not a polygon.** A balloon is either round or oval, but it does not have straight sides with vertices.

222. **Regular octagon.** A stop sign is an 8-sided polygon with all sides of equal length.

223. **Not a polygon.** Like the human face, lace is very intricate. Unlike the human face, lace has lots of line segments that meet at lots of different points.

Set 45

224. **Hexagon ABCDOE.** As long as you list the vertices in consecutive order, any one of these names will do: BCDOEA, CDOEAB, DOEABC, OEABCD, EABCDO. **Hexagon ABCDOE is concave because the measurement of vertex O exceeds 180°.**

225. Three diagonals can be drawn from vertex O: $\overline{OA}$, $\overline{OB}$, $\overline{OC}$. $\overline{OD}$ and $\overline{OE}$ are not diagonals; they are sides.

226. 720°. Three diagonals can be drawn from vertex O, so there are four triangles created. Therefore, the sum of the interior angles of this 6-sided polygon is 180(4) = 720°.

Set 46

227. Heptagon ORLMNPQ. As long as you list their vertices in consecutive order, any one of these names will do: RLMNPQO, MNPQOLR, NPQOLRM, PQOLRMN, QOLRMNP. **Heptagon ORLMNPQ is concave because vertex N exceeds 180°.**

228. Four diagonals can be drawn from vertex O: $\overline{OL}$, $\overline{OM}$, $\overline{ON}$, $\overline{OP}$.

229. 900°. Four diagonals can be drawn from vertex O, so there are five triangles created. Therefore, the sum of the interior angles of this 7-sided polygon is 180(5) = 900°.

Set 47

230. Octagon SUVWXOYZ. If you list every vertex in consecutive order, then your name for the polygon given is correct. **Also, octagon SUVWXOYZ is concave.** The measures of vertices U, W, O and Z exceed 180°.

231. Five diagonals can be drawn from vertex O: $\overline{OZ}$, $\overline{OS}$, $\overline{OU}$, $\overline{OV}$, $\overline{OW}$.

232. 1080°. Five diagonals can be drawn from vertex O, so there are six triangles created. Therefore, the sum of the interior angles of this 8-sided polygon is 180(6) = 1080°.

Set 48

233. Nonagon HIJKLMNOP. List every vertex in consecutive order and your answer is correct. **Also, nongon HIJKLMNOP is regular and convex.**

234. Six diagonals can be drawn from vertex O: $\overline{OH}$, $\overline{OI}$, $\overline{OJ}$, $\overline{OK}$, $\overline{OL}$, and $\overline{OM}$.

235. **1260°.** Six diagonals can be drawn from vertex O, so there are seven triangles created. Therefore, the sum of the interior angles of this 9-sided polygon is 180(7) = 1260°.

Set 49

236. **Adjacent sides.** Draw polygon CDEFG to see that yes, $\overline{CD}$ and $\overline{DE}$ are next to each other and share an angle.

237. **Diagonals.** When a line segment connects non-adjacent endpoints in a polygon, it is a diagonal.

238. **∠GFE or ∠F.**

239. **Adjacent angles.** Look back at the drawing you made of polygon CDEFG. You can see that ∠E and ∠F are next to one another, separated only by one side of the polygon.

Set 50

For solutions to 240 and 241, refer to image below.

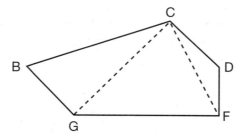

240. **At any one time, three triangles can be drawn in polygon BCDFG.** This is because only two non-overlapping diagonals can be drawn in the polygon. Remember when drawing your triangles that a diagonal must go from endpoint to endpoint.

241. **The interior angles of a convex pentagon will always measure 540° together.** If the interior angles of a triangle measure 180° together, then three sets of interior angles measure 180° × 3, or

540°. Apply the formula $s = 180°(n - 2)$. $s = 180°(5 - 2)$. $s = 180°(3)$. $s = 540°$.

For solutions to 242 and 243, refer to the image below.

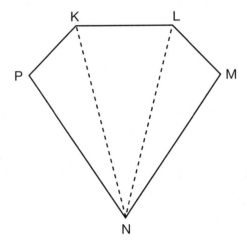

242. At any one time, three triangles can be drawn in polygon **KLMNP. This is because only two non-overlapping diagonals can be drawn.**

243. **$180° \times 3 = 540°$.** Apply the formula $s = 180°(n - 2)$. Again, $s = 540°$. You have again confirmed that the interior angles of a convex pentagon will always measure 540° together.

For solutions to 244 and 245, refer to the image below.

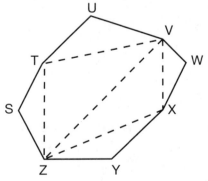

244. At any one time, six triangles can be drawn in polygon **STUVWXYZ.**

245. **$180° \times 6 = 1080°$.** Apply the formula $s = 180°(n - 2)$. $s = 180°(8 - 2)$. $s = 180°(6)$. $s = 1,080°$.

11

Quadrilaterals

As you learned in the last chapter, quadrilaterals are four-sided polygons. The interior of a quadrilateral contains two unique triangles, so the sum of the interior angles of any quadrilateral is 360°. Quadrilaterals are a unique family of polygons because there are so many distinct types of quadrilaterals, such as squares, rectangles, parallelograms, and rhombuses. What's the difference between all of these types of quadrilaterals and how are they classified? Quadrilaterals are divided into three major categories, based on how many pairs of opposite parallel sides they have:

1. The first group of quadrilaterals has no pairs of parallel sides. These quadrilaterals do not have a special name.

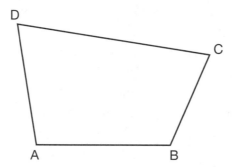

2. The second group of quadrilaterals has exactly one pair of parallel sides and are called **trapezoids**. If the non-parallel opposite sides are congruent, it is called an isosceles **trapezoid**, as pictured in the following illustration.

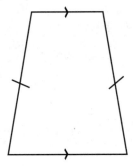

3. The third group of quadrilaterals is **parallelograms**, which have two pairs of parallel sides. A parallelogram is typically drawn as follows, but as we will explore, there are three more specific ways that we can classify parallelograms.

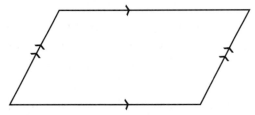

- A **rectangle** is a parallelogram that has four right angles.
- A **square** is a parallelogram that has four right angles and four equal sides. (All squares are rectangles, but not all rectangles are squares!)
- A **rhombus** is any quadrilateral that has four equal sides. All rhombuses are parallelograms. (All squares are rhombuses, but not all rhombuses are squares!)

Each of the special types of quadrilaterals listed above has a list of unique characteristics that pertain to the relationship between the quadrilateral's consecutive angles, opposite angles, and/or diagonals. Read on for a comprehensive list of the characteristics that define each group of quadrilaterals.

Quadrilateral	Four-sided figure that may or may not have a parallel sides
Trapezoid	One pair of parallel sides
Isosceles Trapezoid	One pair of parallel sides Base angles are congruent *Congruent legs* *Congruent diagonals*
Parallelogram	Two pairs of parallel sides Opposite sides are congruent Opposite angles are congruent Adjacent angles are supplementary Diagonals bisect each other
Rectangle	Two pairs of parallel sides Opposite sides are congruent *All angles are congruent* Adjacent angles are supplementary Diagonals bisect each other *Diagonals are congruent*
Rhombus	Two pairs of parallel sides *All sides are congruent* Opposite angles are congruent Adjacent angles are supplementary Diagonals bisect each other *Diagonals bisect the angle of a rhombus* *Diagonals form perpendicular lines*
Square	Two pairs of parallel sides *All sides are congruent* *All angles are congruent* Adjacent angles are supplementary Diagonals bisect each other *Diagonals are congruent* *Diagonals bisect the angle of a square* *Diagonals form perpendicular lines*

Set 51

Choose the best answer.

246. A rhombus, a rectangle, and an isosceles trapezoid all have
    a. congruent diagonals.
    b. opposite congruent sides.
    c. interior angles that measure 360°.
    d. opposite congruent angles.

247. Four line segments connected end-to-end will always form
    a. an open figure.
    b. four interior angles that sum to 360°.
    c. a square.
    d. a parallelogram.

248. Square K has vertices that are the midpoints of square R. The lengths of the sides of square R, as compared to the lengths of the sides of square K are
    a. congruent.
    b. half the length.
    c. twice the length.
    d. none of the above.

249. A figure with four sides and diagonals that bisect each angle could be a
    a. rectangle.
    b. rhombus.
    c. parallelogram.
    d. trapezoid.

250. A figure with four sides and diagonals that bisect each other could NOT be a
    a. rectangle.
    b. rhombus.
    c. parallelogram.
    d. trapezoid.

Use the following figures to answer questions 251 through 254. The figures are not drawn to scale and only the markings on the sides and angles should be used to determine the most specific name you can give to each four-sided polygon.

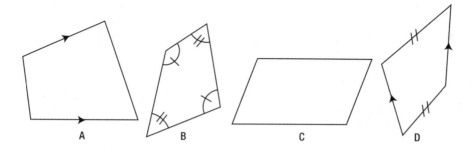

251. Shape A is a(n) _____

252. Shape B is a(n) _____

253. Shape C is a(n) _____

254. Shape D is a(n) _____

Set 52

Fill in the blanks based on your knowledge of quadrilaterals. More than one answer may be correct, so fill in every possibility.

255. If quadrilateral ABCD has two sets of parallel lines, it could be
_____.

256. If quadrilateral ABCD has four congruent sides, it could be
_____.

257. If quadrilateral ABCD has exactly one set of opposite congruent sides, it could be _____.

258. If quadrilateral ABCD has opposite congruent angles, it could be
_____.

259. If quadrilateral ABCD has consecutive angles that are supplementary, it could be _____.

260. If quadrilateral ABCD has congruent diagonals, it could be _____.

261. If quadrilateral ABCD can be divided into two congruent triangles, it could be _____.

262. If quadrilateral ABCD has diagonals that bisect each vertex angle in two congruent angles, it is _____.

Set 53

Choose the best answer.

263. If an angle in a rhombus measures 21°, then the other three angles consecutively measure
   a. 159°, 21°, 159°
   b. 21°, 159°, 159°
   c. 69°, 21°, 69°
   d. 21°, 69°, 69°
   e. It cannot be determined.

264. In an isosceles trapezoid, the angle opposite an angle that measures 62° measures
   a. 62°.
   b. 28°.
   c. 118°.
   d. 180°.
   e. It cannot be determined.

265. In rectangle WXYZ, ∠WXZ and ∠XZY
   a. are congruent.
   b. are alternate interior angles.
   c. form complementary angles with ∠WZX and ∠YXZ.
   d. all of the above
   e. It cannot be determined.

266. In square ABCD, ∠ABD

    a. measures 45°.

    b. is congruent with ∠ADC.

    c. forms a supplementary pair with ∠ADB.

    d. all of the above

    e. It cannot be determined.

267. In parallelogram KLMN, if diagonal KM measures 30 inches, then

    a. $\overline{KL}$ measures 18 inches.

    b. $\overline{LM}$ measures 24 inches.

    c. diagonal LN is perpendicular to diagonal KM.

    d. all of the above

    e. It cannot be determined.

Set 54

Use the figure below to answer questions 268 through 270.

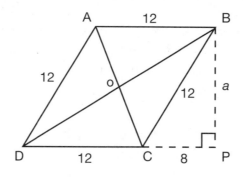

$$m\angle BCA = 72$$
$$m\angle BDA = 18$$

268. Using your knowledge that a rhombus has two pairs of parallel sides, show that diagonals AC and BD intersect perpendicularly. (You cannot just state the a rhombus has perpendicular diagonals.)

269. Using your knowledge of triangles and quadrilaterals, what is the length of imaginary side BP?

270. Using your knowledge of triangles and quadrilaterals, what is the length of diagonal DB?

Answers

Set 51

246. **c.** Rectangles and rhombuses have very little in common with isosceles trapezoids except one set of parallel lines, one set of opposite congruent sides, and four interior angles that measure 360°.

247. **b.** The interior angles of a quadrilateral total 360°. Choices **a** and **c** are incorrect because the question states each line segment connects end-to-end; this is a closed figure, but it is not necessarily a square.

248. **d.** Looking at the illustration, ABCD represents square K and WXYZ represents square R. Since X and W are midpoints, it stands that $\overline{AX} \cong \overline{AW}$. However, the only way that the length of $\overline{WX}$ could be half of $\overline{AB}$ is if $\overline{WX}$ was also congruent to $\overline{AX}$ and $\overline{AW}$. This could only be the case if ΔAWX were equilateral, but it is an isosceles right triangle. Therefore the side lengths of square R are not half of the side lengths of square K. They are definitely not congruent or twice as long as the sides of square K, so the correct choice is **d.**

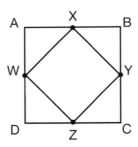

249. **b.** A rhombus's diagonal bisects its vertices.

250. **d.** Diagonals of a trapezoid are not congruent unless the trapezoid is an isosceles trapezoid. Diagonals of any trapezoid do not bisect each other.

251. **Trapezoid.** Trapezoids have exactly one pair of parallel sides.

252. Parallelogram. Parallelograms have two pairs of opposite, congruent angles.

253. Quadrilateral. The polygon *looks* like a parallelogram, but has no markings to indicate congruence or parallel sides, so it can only be classified as a quadrilateral.

254. Isosceles trapezoid. Isosceles trapezoids have one pair or parallel sides and one pair of opposite congruent sides.

Set 52

255. A parallelogram, a rectangle, a rhombus, or a square. Two pairs of parallel lines define each of these four-sided figures.

256. A rhombus or a square.

257. An isosceles trapezoid.

258. A parallelogram, a rectangle, a rhombus, or a square. When a transversal crosses a pair of parallel lines, alternate interior angles are congruent, while same side interior angles are supplementary. Draw a parallelogram, a rectangle, a rhombus, and a square; extend each of their sides. Find the "Z" and "C" shaped intersections in each drawing.

259. A parallelogram, a rectangle, a rhombus, or a square. Again, look at the drawing you made above to see why consecutive angles are supplementary.

260. A rectangle, a square, or an isosceles trapezoid.

261. A parallelogram, a rectangle, a rhombus, or a square.

262. A rhombus or a square.

Set 53

263. a. The first consecutive angle must be supplementary to the given angle. The angle opposite the given angle must be congruent.

Consequently, in consecutive order, the angles measure $180° - 21°$, or $159°$, $21°$, and $159°$. Choice **b** does not align the angles in consecutive order; choice **c** mistakenly subtracts $21°$ from $90°$ when consecutive angles are supplementary, not complementary.

264. **c.** Opposite angles in an isosceles trapezoid are supplementary. Choice **a** describes a consecutive angle along the same parallel line.

265. **d.** $\overline{XZ}$ is a diagonal in rectangle WXYZ. $\angle WXZ$ and $\angle XZY$ are alternate interior angles along the diagonal; they are congruent; and when they are added with their adjacent angle, the two angles form a $90°$ angle.

266. **a.** $\overline{BD}$ is a diagonal in square ABCD. It bisects vertices B and D, creating four congruent $45°$ angles. Choice **b** is incorrect because $\angle ABD$ is half of $\angle ADC$; they are not congruent. Choice **c** is incorrect because when two $45°$ angles are added together they measure $90°$, not $180°$.

267. **e.** It cannot be determined.

Set 54

268. First, opposite sides of a rhombus are parallel, which means alternate interior angles are congruent. If $\angle BCA$ measurements $72°$, then $\angle CAD$ also measures $72°$. The sum of the measurements of all three interior angles of a triangle must equal $180°$: $72° + 18° + m\angle AOB = 180°$. $m\angle AOD = 90°$. Because $\overline{AC}$ and $\overline{DB}$ are intersecting straight lines, if one angle of intersection measures $90°$, all four angles of intersection measure $90°$, which means the lines perpendicularly meet.

269. $a = 4\sqrt{5}$. $\overline{BP}$ is the height of rhombus ABCD and the leg of $\triangle BPC$. Use the Pythagorean theorem: $a^2 + 8^2 = 12^2$. $a^2 + 64 = 144$. $a^2 = 80$. $a = 4\sqrt{5}$.

270. $c = 4\sqrt{30}$. Use the Pythagorean theorem to find the hypotenuse of $\triangle BPD$, which is diagonal BD: $(4\sqrt{5})^2 + (12 + 8)^2 = c^2$. $80 + 400 = c^2$. $480 = c^2$. $4\sqrt{30} = c$.

12

Perimeter of Polygons

Many professionals work with perimeter on a daily basis. The *perimeter* is the total distance around the outer edge of a shape. In some professions, this distance is needed in order to purchase the correct amount of supplies for a project. In *irregular* polygons, simply add up the lengths of all of the sides in order to calculate a figure's perimeter. *Regular* polygons have n equal sides that are all s units long. Therefore, the perimeter can be found by multiplying the number of sides by the length of each side: $P = ns$. The following illustration shows a few other types polygons that have perimeter formulas.

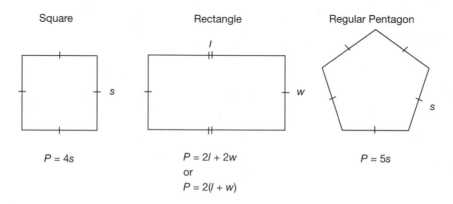

Square

$P = 4s$

Rectangle

$P = 2l + 2w$
or
$P = 2(l + w)$

Regular Pentagon

$P = 5s$

Set 55

Choose the best answer.

271. A regular octagonal gazebo is added to a Victorian lawn garden. Each side of the octagon measures 5 ft. The formula for the gazebo's perimeter is
  a. $p = 8 \times 5$.
  b. $8 = n \times 5$.
  c. $5 = n \times 8$.
  d. $s = n \times p$.

272. The perimeter of a rectangular volleyball court is 54 meters. The length of the court is 18 meters. What is the width of the court?
  a. 36 meters
  b. 18 meters
  c. 9 meters
  d. cannot be determined

273. The perimeter of a square plot of property is 1,600 ft. Each side measures
  a. 4 ft.
  b. 40 ft.
  c. 400 ft.
  d. 4,000 ft.

274. Roberta draws two similar, but irregular pentagons. The perimeter of the larger pentagon is 93 ft.; one of its sides measures 24 ft. If the perimeter of the smaller pentagon equals 31 ft., then the corresponding side of the smaller pentagon measures
   a. $5s = 31$.
   b. $93s = 24 \times 31$.
   c. $93 \times 24 = 31s$.
   d. $5 \times 31 = s$.

275. The deer are eating all of Uncle Bill's vegetables from his garden! Aunt Mary finds a roll of mesh fencing in the garage, which contains 72 feet of fencing. They would like to use all of the fencing to enclose a rectangular garden space that is twice as long as it is wide. What should the dimensions of the garden be?
   a. 14 meters by 28 meters
   b. 14 meters by 22 meters
   c. 18 meters by 18 meters
   d. 12 meters by 24 meters

Set 56

Choose the best answer.

276. Which perimeter is not the same?

a.

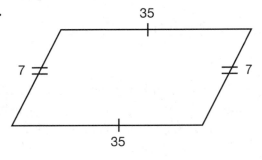

b.

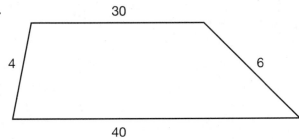

c.

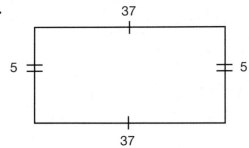

d.

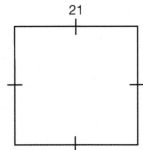

277. Which perimeter is not the same?
  a. a 12-foot square backyard
  b. an 8-foot regular hexagon pool
  c. a 6-foot regular octagonal patio
  d. a 4-foot regular decagon Jacuzzi
  e. It cannot be determined.

278. Rectangle R has a perimeter that is twice as large as the perimeter of square S. The length of rectangle R is three times as long as a side of square S. If the length of a side of square S is s, what algebraic expression would express the width of rectangle R?
  a. s
  b. $2s$
  c. $3s$
  d. $4s$

279. The measure of which figure's side is different from the other four figures?
  a. a regular nonagon whose perimeter measures 90 feet
  b. an equilateral triangle whose perimeter measures 27 feet
  c. a regular heptagon whose perimeter measures 63 feet
  d. a regular octagon whose perimeter measures 72 feet
  e. It cannot be determined.

280. Which figure does not have 12 sides?
  a. Regular Figure A with sides that measure 4.2 in. and a perimeter of 50.4 in.
  b. Regular Figure B with sides that measure 1.1 in. and a perimeter of 13.2 in.
  c. Regular Figure C with sides that measure 5.1 in. and a perimeter of 66.3 in.
  d. Regular Figure D with sides that measure 6.0 in. and a perimeter of 72.0 in.
  e. It cannot be determined.

Set 57

Find the perimeter of the following figures.

281.

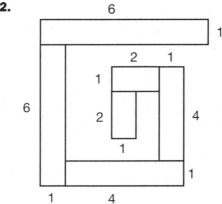

P = _____

282.

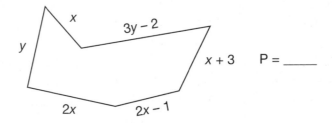

283.

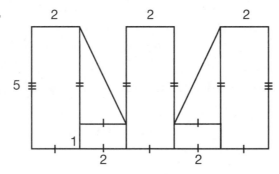

284.

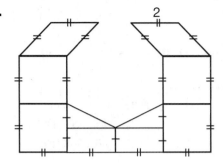

Set 58

Use the figure below to answer questions 285 through 286.

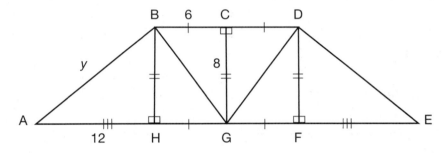

285. Find the value of *y*.

286. Find the figure's total perimeter.

Set 59

Use the figure below to answer questions 287 through 288.

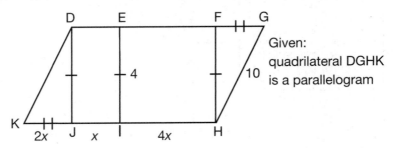

Given:
quadrilateral DGHK
is a parallelogram

287. Find the value of x.

288. Find the figure's total perimeter.

Set 60

Use the figure below to answer questions 289 through 291.

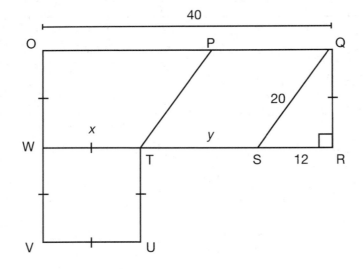

Given:
$\overline{OQ} \cong \overline{WR}$
$\overline{PQ} \cong \overline{TS}$

289. Find the value of x.

290. Find the value of y.

291. Find the figure's total perimeter.

Set 61

Use the figure below to answer questions 292 through 294.

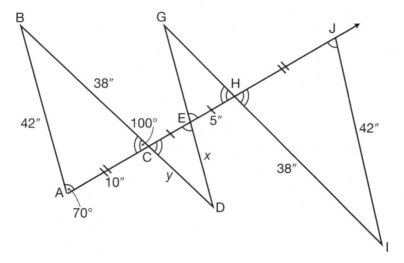

292. Find the value of x.

293. Find the value of y.

294. Find the figure's total perimeter.

Answers

Set 55

271. **a.** To find the perimeter, multiply the number of sides by the measure of one side. Octagons have eight sides and the question states that each side measures 5 feet. The perimeter of this Victorian gazebo is $p = 8 \times 5$.

272. **c.** Work backwards by putting the given information into the perimeter formula for rectangles and then solve for w:

$P = 2l + 2w$
$54 = 2(18) + 2w$
$54 = 36 + 2w$
$18 = 2w$
$w = 9$ meters

273. **c.** Plug the numbers into the formula: $p = ns$. $1600 = 4s$. $400 = s$.

274. **b.** A proportion can find an unknown side of a figure by using known sides of a similar figure; a proportion can also find an unknown side using known perimeters. $\frac{93}{24} = \frac{31}{s}$. Cross-multiply: $93s = 24 \times 31$.

275. **d.** Since they want the garden to be twice as long as it is wide, let the width $= x$ and the length $= 2x$. Then work backwards by putting the width $= x$ and the length $= 2x$ into the perimeter formula for rectangles and then solve for x:

$P = 2l + 2w$
$72 = 2(2x) + 2(x)$
$72 = 4x + 2x$
$72 = 6x$
$x = 12$ meters, so the width will be 12 meters and the length will be 24 meters.

Set 56

276. **b.** Each figure except trapezoid B has a perimeter of 84 feet; its perimeter measures only 80 feet.

277. **d.** Apply the formula $p = ns$ to each choice. In choice **a**, the perimeter of the backyard measures 12 feet × 4 sides, or 48 feet. In choice **b**, the perimeter of the pool measures 8 feet × 6 sides, or 48 feet. In choice **c**, the perimeter of the patio measures 6 feet × 8 sides, or 48 feet. In choice **d**, the perimeter of the Jacuzzi measures 4 feet by 10 sides, or 40. It is obvious that the Jacuzzi has a different perimeter.

278. **a.** Since Square S has four sides that are s long, its perimeter is $4s$. Since Rectangle R has a perimeter that is twice as large as square S, the perimeter of rectangle R is $(4s)(2) = 8s$. It is also known the that the length of rectangle R is $3s$ so put these things into the perimeter formula and solve for width:

$P = 2l + 2w$

$8s = 2(3s) + 2w$

$8s = 6s + 2w$

$2s = 2w$

$w = s$

279. **a.** To find the measure of each side, change the formula $p = ns$ to $\frac{p}{n} = s$. Plug each choice into this formula. In choice **a**, the sides of the nonagon measure $\frac{90 \text{ feet}}{9 \text{ sides}}$, or 10 feet per side. In choice **b**, the sides of the triangle measure $\frac{27 \text{ feet}}{3 \text{ sides}}$, or 9 feet per side. In choice **c**, the sides of the heptagon measure $\frac{63 \text{ feet}}{7 \text{ sides}}$, or 9 feet per side. In choice **d**, the sides of the octagon measure $\frac{72 \text{ feet}}{8 \text{ sides}}$, or 9 feet per side.

280. **c.** To find the number of sides a figure has, change the formula $p = ns$ to $\frac{p}{s} = n$. Plug each choice into this formula. Choices **a** and **b** have 12 sides. In choice **c**, figure C has 13 sides.

Set 57

281. **$P = 6x + 4y$.** Add all of the sides together and combine like terms: $P = 6x + 4y$; $y + x + (3y - 2) + (x + 3) + (2x - 1) + 2x = 6x + 4y$

282. **$p = 50$.** Using your knowledge of rectangles and their congruent sides, you find the measure of each exterior side not given. To find the perimeter, you add the measure of each exterior side together. $1 + 6 + 1 + 6 + 1 + 4 + 1 + 4 + 1 + 2 + 1 + 2 + 1 + 2 + 1 + 3 + 3 + 5 + 5 = 50$.

283. **$p = 34 + 4\sqrt{5}$.** First, find the hypotenuse of at least one of the two congruent triangles using the Pythagorean theorem: $2^2 + 4^2 = c^2$. $4^2 + 16^2 = c^2$. $20 = c^2$. $2\sqrt{5} = c$. Add the measure of each exterior side together: $2 + 5 + 2 + 2 + 2 + 2 + 2 + 5 + 2 + 2\sqrt{5} + 4 + 2 + 4 + 2\sqrt{5} = 34 + 4\sqrt{5}$.

284. **$p = 32 + 2\sqrt{5}$.** First find the hypotenuse of at least one of the two congruent triangles using the Pythagorean theorem: $1^2 + 2^2 = c^2$. $1 + 4 = c^2$. $\sqrt{5} = c$. Add the measure of each exterior side together. $2 + 2 + 2 + 2 + 2 + 2 + 2 + 2 + 2 + 2 + 2 + 2 + 2 + 2 + 2 + 2 + \sqrt{5} + \sqrt{5} = 32 + 2\sqrt{5}$.

Set 58

285. **$y = 4\sqrt{13}$.** $\overline{CG}$ and $\overline{BH}$ are congruent because the opposite sides of a rectangle are congruent. Plug the measurements of $\triangle ABH$ into the Pythagorean theorem: $12^2 + 8^2 = y^2$. $144 + 64 = y^2$. $208 = y^2$. $4\sqrt{13} = y$.

286. **$p = 48 + 8\sqrt{13}$.** Figure ABDE is an isosceles trapezoid; $\overline{AB}$ is congruent to $\overline{ED}$. Add the measurements of each exterior line segment together: $6 + 6 + 4\sqrt{13} + 12 + 6 + 6 + 12 + 4\sqrt{13} = 48 + 8\sqrt{13}$.

Set 59

287. $x = \sqrt{21}$. In parallelogram DGHK, opposite sides are congruent, so ΔKDJ and ΔGHF are also congruent (Side-Side-Side postulate or Side-Angle-Side postulate). Plug the measurements of ΔKDJ and ΔGHF into the Pythagorean theorem: $(2x)^2 + 4^2 = 10^2$. $4x^2 + 16 = 100$. $4x^2 = 84$. $x^2 = 21$. $x = \sqrt{21}$.

288. $p = 14\sqrt{21} + 20$. Replace each x with $\sqrt{21}$ and add the exterior line segments together: $2\sqrt{21} + \sqrt{21} + 4\sqrt{21} + 10 + 2\sqrt{21} + 4\sqrt{21} + \sqrt{21} + 10 = 14\sqrt{21} + 20$.

Set 60

289. $x = 16$. The hatch marks indicate that $\overline{WT}$ and $\overline{QR}$ are congruent. Plug the measurements of ΔSQR into the Pythagorean theorem: $12^2 + x^2 = 20^2$. $144 + x^2 = 400$. $x^2 = 256$. $x = 16$.

290. $y = 12$. Opposite sides of a rectangle are congruent. $\overline{OQ}$ equals the sum of $\overline{WT}$, $\overline{TS}$, and $\overline{SR}$. Create the equation: $40 = 16 + y + 12$. $40 = 28 + y$. $12 = y$.

291. $p = 144$. Add the measure of each exterior line segment together: $40 + 16 + 12 + 12 + 16 + 16 + 16 + 16 = 144$

Set 61

292. *x* = **21 inches.** ΔABC and ΔJIH are congruent (Side-Side-Side postulate). ΔEDC and ΔEGH are also congruent because three angles and a side are congruent. However, ΔABC and ΔJIH are only similar to ΔEDC and ΔEGH (Angle-Angle postulate). A comparison of side $\overline{AC}$ to side $\overline{EC}$ reveals a 10:5 or 2:1 ratio between similar triangles. If $\overline{AB}$ measures 42 inches, then corresponding line segment $\overline{ED}$ measures half as much, or 21 inches.

293. *y* = **19.** Using the same ratio determined above, if $\overline{BC}$ measures 38 inches, then corresponding line segment $\overline{DC}$ measures half as much, or 19 inches.

294. *p* = **270 inches.** Add the measure of each exterior line segment together: $2(42 + 38 + 10) + 2(21 + 19 + 5) = 180 + 90 = 270$ inches.

13

Area of Polygons

Perimeter is the distance around an object. In this chapter you'll work with *area*, which is the amount of surface covered by an object. For example, the number of tiles on a kitchen floor would be found by using an area formula, while the amount of baseboard used to surround the room would be found by using a perimeter formula. Perimeter is always expressed in linear units. Area is always expressed in square units.

If the perimeter of a figure is the outline of a figure, then the **area** of a figure is what is inside the outline; area is the amount of two-dimensional space that a planar figure occupies.

As you'll see on the next page, area is measured in two-dimensional squares. Therefore, the notation for area always includes an exponent of two to indicate two-dimensional space. For example, 40 square meters means 40 squares that are one meter by one meter. This is written as $40m^2$.

Now that you see how area is measured in smaller square units, let's look at the formulas used to calculate area in different polygons.

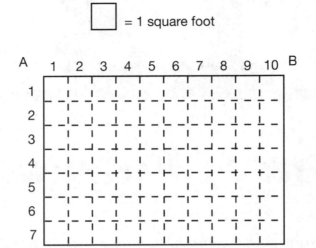

A square equals 1 foot by 1 foot
The area of rectangle ABCD equals 10 squares by 7 squares,
or 70 square feet

Squares

Since the length and width in squares are the same size, it is typical to refer to both of these dimensions as *s*. The formula for the area of a square is the length of one of its sides times itself.

Area of a Square = (side) × (side), or $A = s^2$

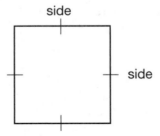

Rectangles

The dimensions of a rectangle are referred to as the *length* and *width*. The formula for the area of a rectangle is length times width.

Area of a Rectangle = (length) × (width), or *A = lw*

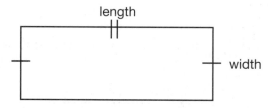

Triangles

The following rectangle has been divided into two equal triangles by a diagonal. Since the area of the rectangle is length times width, you can probably guess that the area of one of the triangles would be half of length times width. With triangles, the names "length" and "width" are not commonly used: side $\overline{BC}$ is referred to as the *base* and side $\overline{DC}$ is the *height*. The formula for the area of a triangle is therefore half of the base times height. A very important thing to remember is that the height must always be at a 90° angle to the base. Sometimes it is necessary to use the Pythagorean theorem to solve for the height, before the area can be calculated.

Area of a Triangle = $\frac{1}{2}$ × (base) × (height) or $A = (\frac{1}{2})bh$

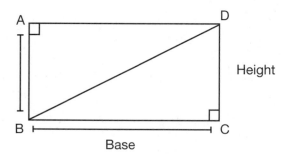

Parallelograms

The dimensions of a parallelogram are referred to as the *base* and *height* as drawn into the figure below. Notice that the height of $\overline{BX}$ is not the length of one of the sides, but the length of the perpendicular line segment connecting the bottom base to the top base. This is always the case with the height of parallelograms. The formula for the area of a parallelogram is base times height.

Area of a Parallelogram = (base) × (height), or *A = bh*

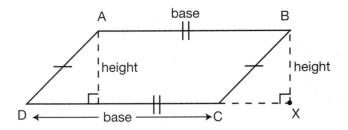

Trapezoids

Trapezoids have two different *bases* and a *height* that is drawn in the same manner as the height of a parallelogram. The height connects the two bases perpendicularly. The formula for the area of a trapezoid is one-half the height times the sum of the two bases.

Area of a Trapezoid = $\frac{1}{2}$(height) × (base$_1$ + base$_2$),
or $A = \frac{1}{2}(h) \times (b_1 + b_2)$

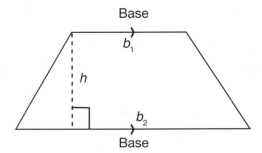

Rhombuses

The area for rhombuses is completely different from the other formulas presented so far. The formula for the area of a rhombus is the product of its two distinct diagonals.

$$\text{\textit{Area of a Rhombus}} = \tfrac{1}{2}(\text{diagonal}_1)(\text{diagonal}_2)$$
$$\text{or } A = \tfrac{1}{2}(d_1)(d_2)$$

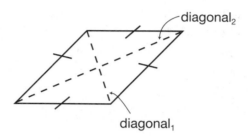

Regular Polygons

Remember that a regular polygon has all equal sides and all equal angles. All regular polygons contain **apothems**. An apothem is a line segment drawn at a right angle from the midpoint of any side to the center of the polygon. All apothems are perpendicular bisectors. The formula for the area of a regular polygon is one-half of the apothem times the perimeter of the polygon. (Remember that the perimeter of regular polygons is ns, where n is the number of sides, and s is the side length.)

$$\text{\textit{Area of a Regular Polygon}} = \tfrac{1}{2} \times \text{\textit{(apothem)(perimeter)}},$$
$$\text{or } A = \tfrac{1}{2} \times (a)(n)(s)$$

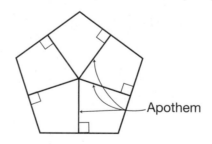

Set 62

Choose the best answer.

295. Area is
   a. the negative space inside a polygon.
   b. a positive number representing the interior space of a polygon.
   c. all the space on a plane.
   d. no space at all.

296. Area is always expressed
   a. in linear units
   b. in metric units
   c. in square units
   d. it depends on the type of polygon adjacent

297. The area of the figure below is the sum of which areas?

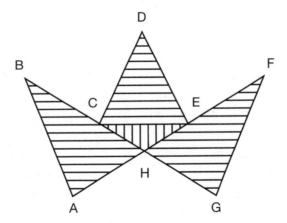

   a. ΔABH + CDEH + ΔHFG + ΔCEH
   b. ΔABH + ΔCDE + ΔHFG
   c. ΔABH + ΔCDE + ΔHFG + ΔCEH
   d. ΔABH + CDEH + ΔHFG + ΔAHG

298. The *height* of a triangle is always
   a. equal to the apothem
   b. equal to the width of a length of the rectangle into which it is inscribed
   c. half of the base of the triangle
   d. a line segment from the base to a vertex, forming a 90-degree angle

299. An apothem
   a. extends from the opposite side of a polygon.
   b. bisects the side of a polygon to which it is drawn.
   c. is drawn to a vertex of a polygon.
   d. forms half of a central angle.

Set 63

Circle whether the statements below are true or false.

300. A rhombus with adjacent sides that measure 5 feet has the same area as a square with adjacent sides that measure 5 feet.
True or False.

301. A rectangle with adjacent sides that measure 5 feet and 10 feet has the same area as a parallelogram with adjacent sides that measure 5 feet and 10 feet.   **True or False.**

302. A rectangle with adjacent sides that measure 5 feet and 10 feet has twice the area of a square with adjacent sides that measure 5 feet.
True or False.

303. A parallelogram with adjacent sides that measure 5 feet and 10 feet has twice the area of a rhombus whose height is equal to the height of the parallelogram and whose adjacent sides measure 5 feet.
True or False.

304. A triangle with a base of 10 and a height of 5 has a third the area of a trapezoid with base lengths of 10 and 20 and a height of 5.
True or False.

Set 64

Find the shaded area of each figure below.

305. Find the shaded area of quadrilateral ABCD.

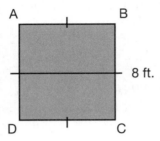

306. Find the shaded area of polygon KLMNO.

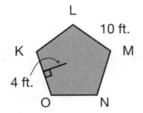

307. Find the shaded area of equilateral ΔDEF.

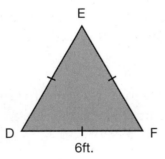

308. Find the shaded area of below in terms of r.

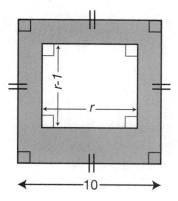

309. Find the shaded area of Figure Y.

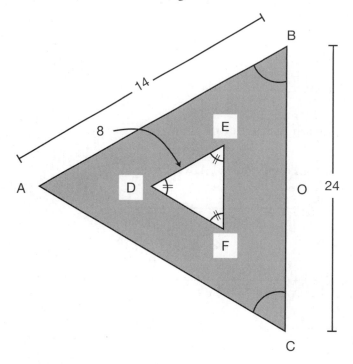

310. Find the area of the shaded figure below.

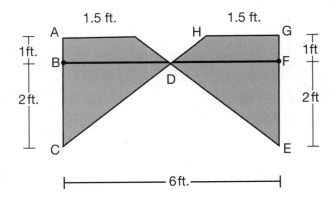

Set 65

Find the area of each figure below.

311. Find the area of quadrilateral ABCD.

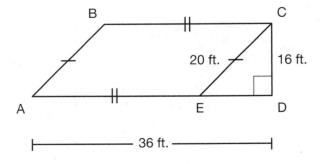

312. Find the area of polygon RSTUV.

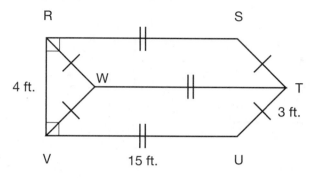

313. Find the area of concave polygon KLMNOPQR.

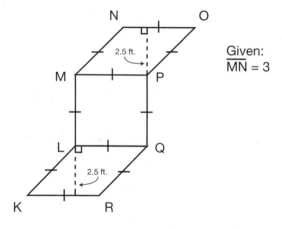

Given:
$\overline{MN} = 3$

314. Find the area of polygon BCDEFGHI.

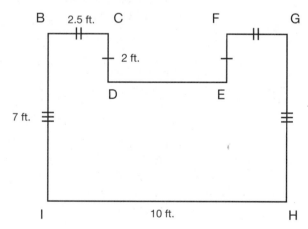

315. Find the area of concave polygon MNOPQR.

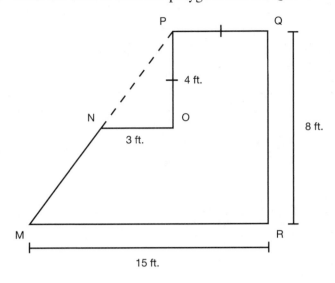

Set 66

Use the figure and information below to answer questions 316 through 319.

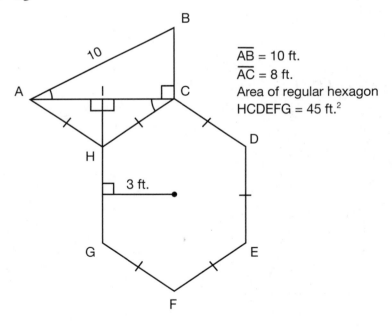

$\overline{AB} = 10$ ft.
$\overline{AC} = 8$ ft.
Area of regular hexagon
HCDEFG = 45 ft.^{2}

316. Find the length of $\overline{DE}$.

317. Find the area of ΔCHI.

318. Find the area of ΔABC.

319. Find the entire area of figure ABCDEFGH.

Set 67

Use the figure and information below to answer questions 320 through 322.

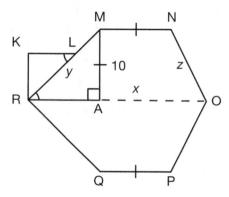

$\overline{RO} = x$
$\overline{RM} = y$
$\overline{NO} = z$
Area of RMNO = Area of RQPO
Area of RMNOPQ = 320 sq. ft.
Area of ΔRMA = 50 sq. ft.

320. Find the measurement of side x.

321. Find the measurement of side y.

322. Find the measurement of side z.

Answers

Set 62

295. **b.** Points, lines, and planes do not occupy space, but figures do. The area of a figure is how much space that figure occupies, always represented by a positive number. Choice **a** is incorrect because area is never a negative number. Choice **c** is incorrect because the area of a plane is infinite; when you measure area, you are only measuring a part of that plane inside a polygon.

296. **c.** Area is a 2-dimensional measurement of space within a planar figure, so it is always expressed in square units, such as four miles^2 to indicate the two dimensions.

297. **c.** The area of a closed figure is equal to the area of its **nonoverlapping** parts. This answer doesn't have to be broken down into all triangles—quadrilateral CDEH is a part of the figure. However, none of the answers can include quadrilateral CDEH and △CEH because they share interior points. Also, △AHG is not part of the closed figure; in fact, it isn't closed at all.

298. **d.** The *height* of a triangle is always a line segment from the base to a vertex, forming a 90-degree angle. The most important thing about height is that it forms a right angle with the base.

299. **b.** An apothem extends from the center of a polygon to a side of the polygon. All apothems are perpendicular bisectors and only span half the length of a polygon. A radius (to be discussed in a later chapter) extends from the center point of a polygon to any vertex. Two consecutive radii form a central angle. Apothems are not radii.

Set 63

300. **False**. The rhombus is not a square, it is a tilted square which makes its height less than 5 feet. Consequently, the area of the square is 25 square feet, but the area of the rhombus is less than 25 square feet.

301. **False**. A parallelogram is not a rectangle. It is a tilted rectangle which makes its height less than 5 feet in this example. Conseqently, the area of the rectangle is 50 square feet, but the area of the parallelogram is less than 50 square feet.

302. **True**. If two squares can fit into one rectangle, then the rectangle has twice the area of one square.

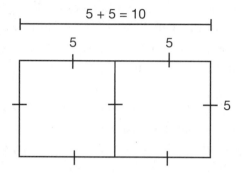

303. **True**. Like the squares and rectangle above, if two rhombuses can fit into one parallelogram, then the parallelogram has twice the area of one rhombus.

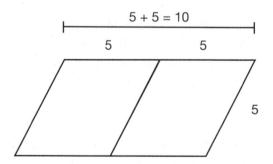

304. **True**. One triangle has an area of 25 square feet. The trapezoid has an area that measures 75 square feet. Three triangles fit into one trapezoid or the area of one triangle is a third of the area of the trapezoid.

Set 64

305. **64 feet².** If one side of the square measures 8 feet, the other three sides of the square each measure 8 feet. Multiply two sides of the square to find the area: 8 feet × 8 feet = 64 square feet.

306. **100 feet².** Sides and angles of a regular polygon are equal so, if one side measures 10 feet, the other sides will also measure 10 feet. The perimeter of this pentagon measures 50 feet (10 × 5 = 50) and its apothem measures 4 feet, so the area of the pentagon measures $\frac{1}{2}$ × 4 feet × 50 feet = 100 square feet.

307. **9√3 feet².** To find the height of equilateral ΔDEF, draw a perpendicular line segment from vertex E to the midpoint of $\overline{DF}$. This line segment divides ΔDEF into two congruent right triangles. Plug the given measurement into the Pythagorean theorem: $(\frac{1}{2} \times 6)^2 + b^2 = 6^2$; $9 + b^2 = 36$; $b = \sqrt{27}$; $b = 3\sqrt{3}$. To find the area, multiply the height by the base: $3\sqrt{3}$ feet × 6 feet = $18\sqrt{3}$ square feet. Then, take half of $18\sqrt{3}$ to get $9\sqrt{3}$.

308. **$100 - r^2 + 1r$**
    The area of the larger square will be 10 × 10 = 100. The area of the smaller contained rectangle will be $A = l \times w = (r)(r - 1) = r^2 - 1r$. To find the area of the shaded region, subtract the smaller figure from the larger: $100 - (r^2 - 1r) = 100 - r^2 + 1r$.

309. **$24\sqrt{13} - 16\sqrt{3}$**
    Solve for the height AO of ΔABC by using leg BO = 12, hypotenuse AB = 14, and the Pythagorean theorem: $14^2 = 12^2 + a^2$, $52 = a^2$, $a = 2\sqrt{13}$. Then $A = \frac{1}{2}(b)(h) = \frac{1}{2}(24)(2\sqrt{13}) = 24\sqrt{13}$, which is the area of ΔABC. Do the same for ΔDEF using a leg of 4 and a hypotenuse of 8: $8^2 = b^2 + 4^2$, $48 = b^2$, $4\sqrt{3} = b$. Then $A = \frac{1}{2}(b)(h) = \frac{1}{2}(8)(4\sqrt{3}) = 16\sqrt{3}$, which it the area of ΔDEF. Subtract the area of ΔDEF from the area of ΔABC to get the shaded area.

310. **10.5 feet².** Find the area of a rectangle with sides 6 feet and 3 feet: A = 6 ft. × 3 ft. = 18 sq. ft. Find the area of both triangular voids: Area of the smaller triangular void = $\frac{1}{2}$(3 ft. × 1 ft.) = 1.5 sq. ft. Area of the larger triangular void = $\frac{1}{2}$(6 ft. × 2 ft.) = 6 sq. ft. So, the total area of the triangular voids is 1.5 ft. + 6 ft. = 7.5 sq. ft. Subtract 7.5 sq. ft. from 18 sq. ft. (the area of the full rectangle) and 10.5 square feet remain.

Set 65

311. **480 feet².** You can either treat figure ABCD like a trapezoid or like a parallelogram and a triangle. However you choose to work with the figure, you must begin by finding the measurement of $\overline{ED}$ using the Pythagorean theorem: $a^2 + 16^2 = 20$. $a^2 + 256 = 400$. $a^2 = 144$. $a = 12$. Subtract 12 feet from 36 feet to find the measurement of $\overline{BC}$: 36 − 12 = 24 feet. Should you choose to treat the figure like the sum of two polygons, find the area of each polygon separately and add them together. Area of parallelogram ABCE: 16 ft. × 24 ft. = Total area of the quadrilateral: 384 sq. ft. Area of ΔECD: $\frac{1}{2}$ × 16 ft. × 12 ft. = 96 sq. ft. Total area of the quadrilateral: 384 sq. ft. + 96 sq. ft. = 480 sq. ft. Should you choose to treat the figure like a trapezoid and need to find the area, simply plug in the appropriate measurements: $\frac{1}{2}$ × 16 ft. (24 ft. + 36 ft.) = 480 feet².

312. **60 + 2√5 feet².** Extend $\overline{TW}$ to $\overline{RV}$. Let's call this $\overline{XW}$. $\overline{XW}$ perpendicularly bisects $\overline{RV}$; as a perpendicular bisector, it divides isosceles triangle RWV into two congruent right triangles and establishes the height for parallelograms RSTW and VUTW. Solve the area of parallelogram VUTW: 2 ft. × 15 ft. = 30 sq. ft. Find the height of ΔRWV using the Pythagorean theorem: $a^2 + 2^2$ = 3^2. $a^2 + 4 = 9$. $a^2 = 5$. $a = \sqrt{5}$. Solve the area of ΔRWV: $\frac{1}{2}$ × $\sqrt{5}$ ft. × 4 ft. = $2\sqrt{5}$ sq. ft. Add all the areas together: $2\sqrt{5}$ sq. ft. + 30 sq. ft. + 30 sq. ft. = 60 + $2\sqrt{5}$ feet².

313. **24.0 feet².** Rhombuses KLQR and MNOP are congruent. Their areas each equal 2.5 ft. × 3 ft. = 7.5 sq. ft. The area of square LMPQ equals the product of two sides: 3 ft. × 3 ft. = 9 ft. The sum of all the areas equal 9 sq. ft. + 7.5 sq. ft. + 7.5 sq. ft. = 24 feet².

314. **60.0 feet².** The simplest way to find the area of polygon BCDEFGHI is to find the area of rectangle BGHI: 10 ft. × 7 ft. = 70 sq. ft. Subtract the area of rectangular void CFED: 5 ft. × 2 ft. = 10 sq. ft. Then subtract the two areas. 70 sq. ft. – 10 sq. ft. = 60 square feet.

315. **70 feet².** Again, the simplest way to the find the area of polygon MNOPQR is to find the area of trapezoid MPQR. $\frac{1}{2}$ × 8 feet (4 ft. + 15 ft.) = $\frac{1}{2}$ × 8(19) = 76 sq. ft. Subtract the area of ΔNPO: $\frac{1}{2}$ × 3 ft. × 4 ft. = 6 sq. ft. 76 sq. ft. – 6 sq ft. = 70 square feet.

Set 66

316. $\overline{\text{DE}}$ **= 5 feet.** To find $\overline{\text{DE}}$, use the given area of hexagon HCDEFG and work backwards. The area of a regular polygon equals half the product of its perimeter by its apothem: 45 sq. ft. = $\frac{1}{2} p$ × 3 ft.; p = 30 ft. The perimeter of a regular polygon equals the length of each side multiplied by the number of sides: 30 ft. = s ft. × 6.; s = 5 ft.

317. **6 feet².** ΔACH is an isosceles triangle. A line drawn from its vertex to $\overline{\text{AC}}$ bisects the line segment, which means $m\overline{\text{AI}} = m\overline{\text{CI}}$, or $\frac{1}{2}$ of 8 feet long. Since question 316 found the measurement of $\overline{\text{HC}}$, only the measurement of $\overline{\text{HI}}$ remains unknown. Plug the given measurements for ΔCHI into the Pythagorean theorem. $4^2 + b^2 = 5^2$. $16 + b^2 = 25$. $b^2 = 9$. $b = 3$. Once the height is established, find the area of ΔCHI: $\frac{1}{2}$ × 4 ft. × 3 ft. = 6 feet².

318. **24 feet².** Using the Pythagorean theorem with ΔABC shows that the height of the triangle is 6: $10^2 = 8^2 + 6^2$. Therefore, $A = \frac{1}{2}bh = \frac{1}{2}(8)(6) = 24$ feet².

319. **81 square feet.** The areas within the entire figure are the sum of its parts: 24 sq. ft. + 6 sq. ft. + 6 sq. ft. + 45 sq. ft. = 81 square feet.

Set 67

320. **22 feet.** The area of trapezoid RMNO plus the area of trapezoid RQPO equals the area of figure RMNOPQ. Since trapezoids RMNO and trapezoid RQPO are congruent, their areas are equal: $\frac{1}{2}$(320 sq. ft.) = 160 sq. ft. The congruent height of each trapezoid is known, and one congruent base length is known (MA $\cong$ MN). Using the equation to find the area of a trapezoid, create the equation: 160 sq. ft. = $\frac{1}{2}$(10 ft.)(10 ft. + x). 160 sq. ft. = 50 sq. ft. + 5x ft. 110 sq. ft. = 5x ft. 22 feet = x.

321. **$10\sqrt{2}$ feet.** Work backwards using the given area of $\triangle$RMA: 50 sq. ft. = $\frac{1}{2}b$(10 ft.). 50 sq. ft. = 5 ft. $\times$ b. 10 ft. = b. Once the base and height of $\triangle$RMA are established, use the Pythagorean theorem to find $\overline{RM}$: $10^2 + 10^2 = c^2$. $100 + 100 = c^2$. $200 = c^2$. $10\sqrt{2} = c$. $\overline{RM} = 10\sqrt{2}$ feet.

322. **$2\sqrt{26}$ feet.** Imagine a perpendicular line from vertex N to the base of trapezoid RMNO. This imaginary line divides $\overline{RO}$ into another 10-foot segment. The remaining portion of line $\overline{RO}$ is 2 feet long. Use the Pythagorean theorem to find the length of $\overline{NO}$: $(10 \text{ ft.})^2 + (2 \text{ ft.})^2 = z^2$. 100 sq. ft. + 4 sq. ft. = z^2. 104 sq. ft. = z^2. $2\sqrt{26}$ feet = z.

14

Surface Area of Prisms

A prism is a polygon in three dimensions. A cube is a square prism that you are probably familiar with. Shoeboxes and pizza boxes are rectangular prisms. Prisms have two congruent polygons as ends, and their sides are made up of parallelograms. The sides of prisms are called **faces**, while the congruent ends are referred to as **bases**. The **surface area** of a prism refers to the collective area of all of the faces and bases of a three-dimensional shape. Surface area can be determined with formulas or by breaking the prism down into the sum of its two-dimensional components.

Surface Area of a Rectangular Prism

A prism has six faces; each face is a rectangle. For every side or face you see, there is a congruent side you cannot see.

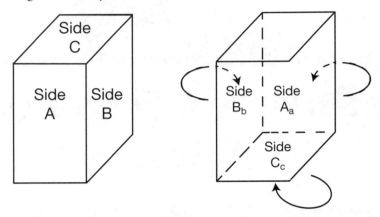

If you pull each face apart, you will see pairs of congruent rectangles.

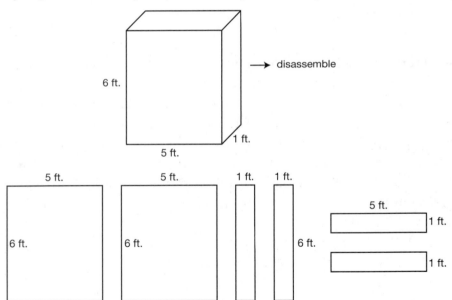

The surface area of a prism is the sum of the areas of its face areas, or $SA = 2(\text{length} \times \text{width}) + 2(\text{width} \times \text{height}) + 2(\text{length} \times \text{height})$. This formula simplifies into:

$$SA = 2(lw + wh + lh)$$

Surface Area of a Cube

Like the rectangular prism, a cube has six faces; each face is a congruent square.

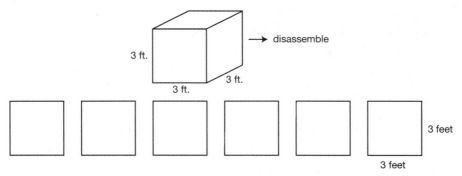

The surface area of a cube is the sum of its face areas, or SA = 6(length × width). This formula simplifies into: $SA = 6e^2$, where e is the measurement of the edge of the cube, or length of one side.

Surface Area of Other Types of Prisms

What about calculating the surface area of three-dimensional prisms that aren't cubes or rectangular prisms? Whether you are dealing with a triangular prism, a pentagonal prism or any other n-sided style of prism, you can always calculate the surface area in parts and then add them together. Remember, every prism is made up of two congruent faces connected by parallelograms. Since you know how to calculate the areas of those shapes, you have the skills to find the surface area of a number of complex prisms. Here are examples of a triangular prism and of a pentagonal prism.

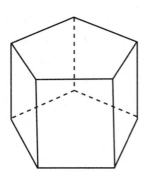
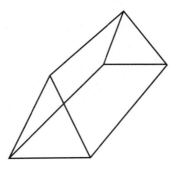

Set 68

Choose the best answer.

323. A rectangular prism has
   a. one set of congruent sides.
   b. two pairs of congruent sides.
   c. three pairs of congruent sides.
   d. four pairs of congruent sides.

324. A cube with sides of length x centimeters has a surface area of $6x^2$ cm^2. If the length of each side of the cube is tripled, what will the surface area be of the resulting cube?
   a. $9x^2$ cm^2
   b. $72x^2$ cm^2
   c. $18x^2$ cm^2
   d. $216x^6$ cm^2

Set 69

Find the surface area.

325. Mark plays a joke on Tom. He removes the bottom from a box of bookmarks. When Tom lifts the box, all the bookmarks fall out. What is the surface area of the empty box Tom is holding if the box measures 5.2 inches long by 17.6 inches high and 3.7 inches deep?

326. Marvin wrapped a gift in a box, which is a perfect cube with sides that each measured d inches. Maya arrives with a gift that is also in a box that is a perfect cube, but the edge length of each side of her gift is twice as long as Marvin's gift. How many square inches of wrapping paper, in terms of d, will Maya need to wrap her gift?
   a. $6d^2$
   b. $12d^2$
   c. $24d^2$
   d. $2d^3$

327. Jimmy gives his father the measurements of a table he wants built. If the drawing below represents that table, how much veneer does Jimmy's father need to buy in order to cover all the exterior surfaces of his son's table? (Include the bottom of the table and the bottom of the legs.)

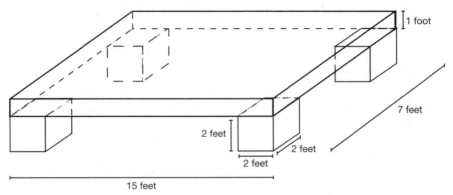

328. In April or May, Victoria will take a troupe of Girl Scouts camping in Oregon. Since there is a good chance that the weather will be wet and rainy, Victoria wants to coat the group's tent with a waterproof liquid coating. One bottle of waterproof coating will cover 100 square feet of fabric. The tent that Victoria is bringing is a triangular prism that is 14 feet long, with a triangular entryway that is 8 feet wide and 5 feet tall. If she wants to coat all of the surfaces, including the bottom, how many bottles of waterproof coating should Victoria buy?

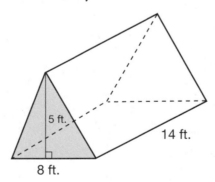

329. Sarah cuts three identical blocks of wood and joins them end-to-end. How much exposed surface area remains (round up to the nearest tenth)?

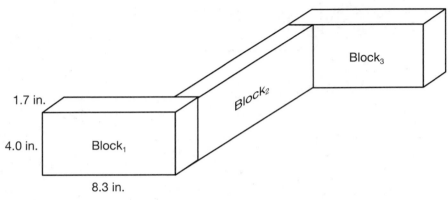

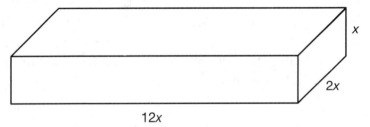

1.7 in.

4.0 in.   Block₁

8.3 in.

SA Block₁ > SA Block₂ > SA Block₃

Set 70

Find each value of *x* using the figures and information below.

330. *Surface Area* = 304 square feet

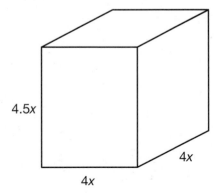

x

$2x$

$12x$

331. *Surface Area* = 936 square meters

$4.5x$

$4x$

$4x$

332. *Surface Area* = 720 square yards

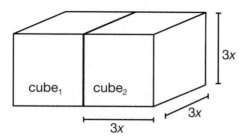

$cube_1 \cong cube_2$

Answers

Set 68

323. **c.** When the faces of a rectangular prism are laid side-by-side, you always have three pairs of congruent faces. That means every face of the prism (and there are six faces) has one other face that shares its shape, size, and area.

324. **b.** The original cube has a side length of x, so after it is tripled, the side length will by $3x$. A cube with a side length of $3x$ will have six faces that each have an area of $9x^2$. Therefore the surface area will be $6(9x^2)$ cm^2 = $72x^2$ cm^2. (Although it is tempting to just triple $6x^2$ cm2 to $18x^2$ cm^2, that will not work because it does not take into consideration that *each* side length has tripled.)

Set 69

325. *Surface Area* = **260.24 square inches.** Begin by finding the whole surface area: surface area = (SA = 2[(17)(5.2) + (5.2)(3.7) + (17.6)(3.7) = 351.76 in^2.) From the total surface area, subtract the area of the missing face: Remaining SA = 351.76 sq. in. – 91.52 sq. in. Remaining SA = 260.24 square inches.

326. **c.** The surface area of Marvin's gift is $6d^2$. Although it is tempting to just double $6d^2$ to $12d^2$, that will not work because it doesn't take into consideration that *each* side length has doubled. Instead you must consider Maya's gift that has an edge length of $2d$, which will in turn have 6 faces that are each $4d^2$ in area. The total surface area of Maya's gift will be $6 \times 4d^2 = 24d^2$.

327. *Surface Area* = **318 feet^2.** These next few problems are tricky: Carefully look at the diagram. Notice that the top of each cubed leg is not an exposed surface area, nor is the space they occupy under the large rectangular prism. Let's find these surface areas first. The top of each cubed leg equals the square of the length of the cube: (2 feet) = 4 sq. ft. There are four congruent cubes, four congruent faces: 4 × 4 sq. ft. = 16 sq. ft. It is reasonable to assume that where the cubes meet the rectangular prism, an equal amount of area from the prism is also not exposed. Total area concealed = 16 sq. ft + 16 sq. ft. = 32 sq. ft. Now find the total surface area of the table's individual parts.

SA of one cube = 6(2 feet)^2 = 6(4 sq. ft.) = 24 sq. ft.

SA of four congruent cubes = 4 × 24 sq. ft. = 96 sq. ft.

SA of one rectangular prism = 2(15 ft.(7 ft.) + 7 ft.(1 foot) + 15 ft.(1 foot)) = 2(105 sq. ft. + 7 sq. ft. + 15 sq. ft.) = 2(127 sq. ft.) = 254 sq. ft.

Total SA = 96 sq. ft. + 254 sq. ft. = 350 sq. ft.

Finally, subtract the concealed surface area from the total surface area = 350 sq. ft. – 32 sq. ft = 318 sq. ft.

328. **4 bottles.** Since this is a triangular prism, break it into it individual parts. First calculate the area of the triangular face, or door of the tent. $A = \frac{1}{2}(b)(h) = \frac{1}{2}(8)(5) = 20$ ft^2. Since there are triangular doors at the front and the back, double this to get 40 ft^2. Next, the bottom of the tent will be a rectangle that is 8 feet by 14 feet, so $A = (w)(l) = (8)(14) = 112$ ft^2. The two remaining sides are a little trickier because the Pythagorean theorem will be needed to solve for the "width" or hypotenuse of the front triangular face: $a^2 + b^2 = c^2$, $4^2 + 5^2 = 16 + 25 = 41 = c^2$, so $c = 6.4$ feet. Use this as the width of the two slanting sides of the tent that have a length of 14. $A = (w)(l) = (6.4)(14) = 89.6$ ft2. Since there are two of these sides, double 89.6ft^2 to get 179.2 ft^2 as the surface area for the two slanted sides of the tent. Add all of the subtotals together to get: 40ft^2 + 112ft^2 + 179.2 ft^2 = 331.2 ft^2. Since each bottle of waterproof coating will just cover 100 square feet, Victoria will need to buy 4 bottles.

329. *Surface Area* **= 297.5 sq. in.** The three blocks are congruent; find the surface area of one block and multiply it by three: SA = 2(8.3 in.(4.0 in.) + 4.0 in.(1.7 in.) + 8.3 in.(1.7 in.)) = 2(33.2 sq. in. + 6.8 sq. in. + 14.11 sq. in.) = 2(54.11 sq. in.) = 108.22 sq. in. 108.22 sq. in. × 3 = 324.66 sq. in. Look at the diagram: The ends of the middle block are concealed, and they conceal an equal amount of space on the other two blocks where they are joined: 2 × 2(4.0 in. (1.7 in.) = 27.2 sq. in. Subtract the concealed surface area from the total surface area: 324.66 sq. in. – 27.2 sq. in. = 297.46 sq. in.

Set 70

330. *x* = **2 feet.** Plug the variables into the formula for the *SA* of a prism: 304 sq. ft. = $2(12x(2x) + 2x(x) + 12x(x))$. 304 sq. ft. = $2(24 x^2 + 2x^2 + 12x^2)$. 304 sq. ft. = $2(38x^2)$. 304 sq. ft. = $76x^2$. 4 sq. ft. = x^2. 2 feet = *x*.

331. *x* = **3 meters.** Plug the variables into the formula for the *SA* of a prism: 936 square meters = $2(4.5x(4x) + 4x(4x) + 4.5x(4x))$. 936 sq. meters = $2(18x^2 + 16x^2 + 18x^2)$. 936 sq. meters = $2(52x^2)$. 936 sq. meters = $104x^2$. 9 sq. meters = x^2. 3 meters = *x*.

332. *x* = $2\sqrt{2}$ **yards.** To find the surface area of the figure, first find the area of one cube face. $A = (b)(h)$, so area of one face = $(3x)(3x) = 9x^2$. There are 10 cube faces in total showing in this figure, so the total surface area = $(10)(9x^2) = 90x^2$. Given that *SA* = 720, solve to find *x*.

$720 = 90x^2$
$8 = x^2$
$x = 2\sqrt{2}$ yards

15

Volume of Prisms and Pyramids

Is the cup half empty or half full? In geometry, it is neither half empty, nor half full; it is half the **volume**.

Volume is the three-dimensional space within a solid three-dimensional figure. Remember from chapter 14 that surface area defines the outer planes of a three-dimensional object. Everything within that object is volume. Volume is what is inside the shapes you see.

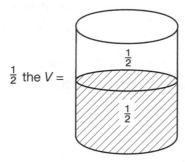

$\frac{1}{2}$ the $V =$

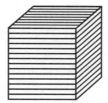

surface area:
the paper it takes
to cover a cube

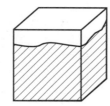

volume:
the liquid it takes to
fill that same cube

Types of Prisms

You investigated rectangular and cubic prisms, among others, in the last chapter. All the three-dimensional polygons you studied in chapter 14 are known as right prisms. The sides of a **right prism** perpendicularly meet the base to form a 90° angle. Remember that the **base** is the polygon that defines the shape of the solid.

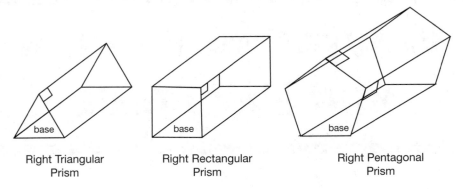

Right Triangular Prism	Right Rectangular Prism	Right Pentagonal Prism

The Volume of a Right Prism

Given *any* right prism, if you can calculate the area of one of its congruent bases, then multiply that area by the height of the prism, you will find its volume. In the following figure, the shaded triangular base on the right side of the prism has an area of $\frac{1}{2}(9)(12) = 54$ cm². If this area is multiplied by the "height" of the prism, which is the 18 cm edge, then that will yield a total volume of 972 cm³. This method can be used with any right prism.

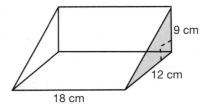

Volume = (area of triangular base) (height)

$V = [(\frac{1}{2})(9)(12)] \times (18)$

$V = 54$ cm³

Rectangles and cubes have special formulas that can be shortcuts:

Volume of a right rectangular prism = (*length*) (*width*) (*height*) = *lwh*

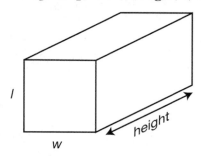

Volume of a right cube = s^3 or e^3
(where *s* or *e* equals the side length)

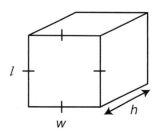

$l = w = h = s = e$
(where *s* equals side and *e* equals edge)

Volume of other right prisms = (*area of base*) (*height*)

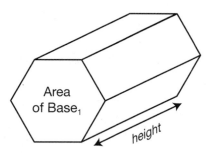

The sides of an **oblique prism** do not meet the base at a 90° angle. Again, that base can be any polygon.

The most common oblique prism is the **pyramid**. A pyramid is a three-dimensional shape with a regular polygon base and congruent triangular sides that meet at a common point. Pyramids are named by the shape of their bases, as demonstrated in the following three figures:

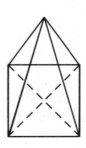
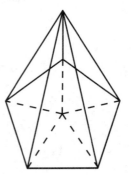

| Triangular Pyramid | Square Pyramid | Pentagonal Pyramid |

The Volume of a Pyramid

The volume of a pyramid is a third of the volume of a right prism that has the same base and height measurements.
The volume of a pyramid = $\frac{1}{3}$ (area of its base) × (height).

The sum of the volumes of three pyramids with square bases will equal the volume of a cube that has the same dimensions:

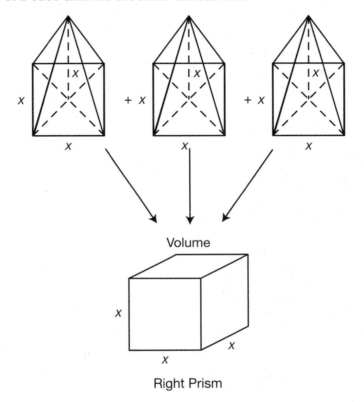

Right Prism

Set 71

Choose the best answer.

333. Which figure below is a right prism?

a.

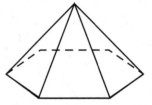

b.

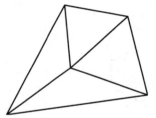

c.

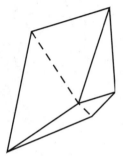

d.

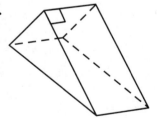

334. Which polygon defines the shape of the right prism below?

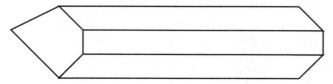

    a. triangle
    b. rectangle
    c. quadrilateral
    d. pentagon

335. What is the surface area of a cube prism that has a volume of 1,000 in^3?
    a. 10 cm^2
    b. 100 cm^2
    c. 600 in^2
    d. 60 in^2

336. Which figure below is a right hexagonal prism?

a.

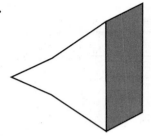

b.

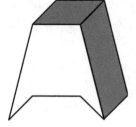

c.

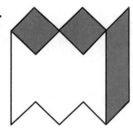

d.

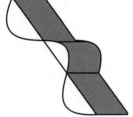

337. Samara is packing her books up after a semester at college. Her roommate gave her an extra box, but it ended up being too small. She went to the market and bought a box that had the same width, but was twice as high and three times as long. If the free box from her roommate only fit seven uniformly sized books, how many books should Samara expect to fit in the larger box she is purchasing if all of her books are about the same size?

   a. 6

   b. 13

   c. 35

   d. 42

338. Which figure below has a third of the volume of a 3 in. cube?

   a.

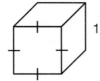

1 in.

   b.

2 in.

   c.

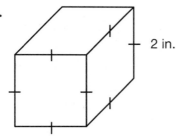

3 in.

   d.

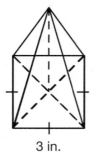

1 in.

339. Miss Sweet has a flower vase that is a regular triangular prism. The vase is a foot and a half tall and its triangular base is a right triangle with legs of four inches and six inches. If Miss Sweet will fill the vase only $\frac{3}{4}$ full with water, how many cubic inches of water will it contain?

    a. 162 in^3

    b. 216 in^3

    c. 432 in^3

    d. 324 in^3

Set 72

Find the volume of each solid.

340. Find the volume of a right heptagonal prism with base sides that measure 13 cm, an apothem that measures 6 cm, and a height that measures 2 cm.

341. Find the volume of a pyramid with four congruent base sides. The length of each base side and the prism's height measure 2.4 ft.

342. Find the volume of a pyramid with an eight-sided base that measures 330 sq. in. and a height that measures 10 in.

Set 73

Find each unknown element using the information below.

343. Find the height of a right rectangular prism with a 295.2 cubic in. volume and a base area that measures 72.0 sq. in.

344. Find the base area of a right nonagon prism with an 8,800 cubic ft. volume and a height that measures 8.8 ft.

345. Find the area of a triangular pyramid's base side if its volume measures $72\sqrt{3}$ cubic meters and its height measures 6 meters.

Set 74

Use the solid figure below to answer questions 346 through 348.

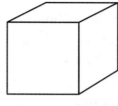

l = 2.1 meters

346. What is the sum of all of the edge lengths in this cube?

347. What is the surface area?

348. What is the volume?

Set 75

Use the solid figure below to answer questions 349 through 351.

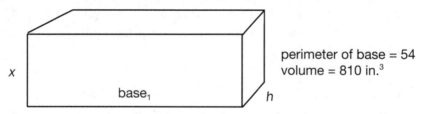

x

base$_1$

h

perimeter of base = 54
volume = 810 in.^{3}

349. Calculate the width and length of one of the figure's biggest sides?

350. What is the height?

351. What is the surface area?

Answers

Set 71

333. **d.** Choice **a** is a hexagonal pyramid; none of its six sides perpendicularly meets its base. The sides of choice **b** only perpendicularly join one base side, and choice **c** is an oblique quadrilateral; its base is facing away from you. Choice **d** is the correct answer; it is a triangular right prism.

334. **d.** The solid in the figure has seven sides. Subtract two base sides, and it has five sides, one for each edge of a pentagon. You will be tempted to answer *rectangle*. Remember all right prisms have rectangles. It is the polygon at the base of the rectangle that defines the prism's shape.

335. **c.** Since the formula for the volume of a right cube = s^3, then 1,000 = s^3, and s = 10 in. The surface area of a cube is 6 times (area of one face). Therefore, the surface area will be $6(10 \times 10) = 600$ in^2.

336. **b.** A hexagonal prism must have a hexagon as one of its sides. A right hexagonal prism has two hexagons. Choice **a** is a pentagonal right prism; choice **c** is a decagonal right prism; and choice **d** is not a prism at all.

337. **d.** Assume that the volume of the first box from Samara's roommate was *lwh*. Since the box Samara purchased had the same width, but was twice as high and three times as long, the new box's volume should be expressed as $(3l)(w)(2h) = 6lwh$. Therefore, the new box has a volume that is six times bigger and it should hold six times as many books. $7 \times 6 = 42$.

338. **c.** Again, you are looking for a pyramid with the same base measurements of the given cube. A 3-inch cube has a volume of 27 in^3. Twenty-seven choice **a**'s can fit into the given cube; meanwhile, eighty-one choice **d**'s fit into that same cube. Only three choice **c**'s fit into the given cube; it has one-third the volume.

339. **a.** Since the triangular base of the vase is a right triangle with legs of 4 and 6 inches, the area of this triangular base is $A = \frac{1}{2}(b)(h) =$ $(\frac{1}{2})(4)(6) = 12$ in². The vase is a foot and a half tall, which means it is 18 inches tall. The total volume of water that the vase can hold is $V = 12 \times 18 = 216$ in³. Since Miss Sweet is only going to fill it $\frac{3}{4}$ full, multiply the full volume of 216 in³ by $\frac{3}{4}$ to get 162 in³.

Set 72

340. *Volume* **= 546 cubic centimeters.** The area of a seven-sided figure equals one-half of its perimeter multiplied by its apothem: *perimeter of heptagonal base* = 13 cm × 7 sides = 91 cm. *Area of heptagonal base* = $\frac{1}{2}$ × 91 cm × 6 cm = 273 square cm. The volume of a right prism is the area of the base multiplied by the prism's height: *volume of prism* = 273 square cm × 2 cm = 546 cubic cm.

341. *Volume* **= 4.6 cubic feet.** A pyramid with four congruent sides means that this is a square based pyramid. Its volume is a third of a cube's volume with the same base measurements, or $\frac{1}{3}$ (area of its base × height). Plug its measurements into the formula: $\frac{1}{3}(2.4$ ft.$)^2$ × 2.4 ft. *Volume of square pyramid* = $\frac{1}{3}$(5.76 sq. ft.) × 2.4 ft. = $\frac{1}{3}$(13.824 cubic ft.) = 4.608 cubic ft.

342. *Volume* **= 1,100 cubic inches.** Unlike the example above, this pyramid has an octagonal base. However, it is still a third of a right octagonal prism with the same base measurements, or $\frac{1}{3}$ (area of its base × height). Conveniently, the area of the base has been given to you: *area of octagonal base* = 330 square inches. *Volume of octagonal pyramid* = $\frac{1}{3}$(330 sq. in) × 10 in. = $\frac{1}{3}$(3,300 cubic in.) = 1,100 cubic in.

Set 73

343. *Height* **= 4.1 inches.** Backsolve by plugging the numbers you are given into the formula for volume of a right rectangular prism.

$V = lwh$
$295.2 = (l)(w)(h)$

You know from chapter 13, page 155, that the area of a rectangle is length times width. So let the given area of 72.0 square inches take the place of *l* and *w* so you can solve for *h*.

$295.2 = 72.0$ sq. in. $\times h$
4.1 in $= h$

344. *Area* **= 1,000 square feet.** Don't let the type of prism, nonagon, fool you. As long as it's a right prism, you can use the formula $V = $ (area of base)(height) to help you. In this question, backsolve to find the area of the base.

$8,800$ cubic ft. $= (b)(8.8$ ft$)$
$1,000$ sq. ft. $= b$

345. *Area* **= $36\sqrt{3}$ = 62.4m^2.** If the volume of a triangular pyramid is $72\sqrt{3}$ cubic meters, work backwards to find the area of its triangular base and then the length of a side of that base (remember, you are working with regular polygons, so the base will be an equilateral triangle). $72\sqrt{3}$ cubic meters $= \frac{1}{3}$ *area of base* $\times$ 6 meters. $72\sqrt{3}$ cubic meters $= a \times 2$ meters. $36\sqrt{3}$ square meters $= a$.

Set 74

346. **25.2 meters.** The perimeter of the front face and back face combined will be double the perimeter of the front face: $P = 4(2.1) = 8.4$, which yields 16.8 meters when doubled. Then there are 4 more edges that are each 2.1 meters long, which sum to 8.4 meters. Together, all these edge lengths sum to 25.2 meters.

347. *Surface area* **= 26.5 square meters.** The surface area of a cube is the area of one face multiplied by the number of faces, or $SA = 6bh$. $SA = 6(2.1 \text{ meters})^2$. $SA = 6(4.41 \text{ square meters})$. $SA = 26.46$ square meters.

348. *Volume* **= 9.3 cubic meters.** The volume of a cube is its length multiplied by its width multiplied by its height, or $V = e^2$ (e represents one edge of a cube). $V = 2.1 \text{ meters} \times 2.1 \text{ meters} \times 2.1 \text{ meters}$. $V = 9.261$ cubic meters.

Set 75

349. *Length* **= 18 inches;** *width* **= 9 inches.** Plug the given variables and perimeter into the formula $p = l + w + l + w$. 54 in. = $2x + x + 2x + x$. 54 in. = $6x$. 9 inches = x.

350. *Height* **= 5 inches.** By now, you remember the formula for a rectangular prism: $V = lwh$. You solved for l and w in question 349 and the volume is given. Backsolve by plugging those numbers into the formula to solve for h.

$810 \text{ in}^3 = (18)(9)(h)$
$810 \text{ in}^3 = (162 \text{ in}^2)(h)$
$5 \text{ inches} = h$

351. *Surface area* **= 594 square inches.** As we learned in chapter 14, the surface area of a prism is a sum of areas, or $SA = 2(lw + wh + lh)$. Plug the measurements you found in the previous questions into this formula. $SA = 2(18 \text{ in.} \times 9 \text{ in.}) + (9 \text{ in.} \times 5 \text{ in.}) + (18 \text{ in.} \times 5 \text{ in.})$. $SA = 2(162 \text{ sq. in.} + 45 \text{ sq. in.} + 90 \text{ sq. in.})$. $SA = 2(297 \text{ sq. in.})$. $SA = 594$ square inches.

16

Introduction to Circles and Arcs

Of course you know *what* a circle looks like, but do you know how it is defined? A **circle** is a collection of points that are an equal distance from one center point. Since circles do not have any corners or vertex points that define their dimensions, there are special formulas, and a special ratio called *pi* (π), that are used when calculating perimeter, area, and volume in circles and spheres.

Center Point, Radius, Central Angle

The center point in a circle is called just that: the **center**! The center point is what is used to name the circle. The following circle would be called *circle C* or ⊙C. The distance from the center of the circle to any point on the edge of the circle is called the **radius** of the circle. The radius is the defining element of a circle, determining how big it is. A circle has an infinite number of **radii**, which are all the same length ("radii" is the plural of "radius"). If two circles share the same center, but have radii of different lengths, then the circles are **concentric**. Below, there are two concentric circles, named C. The smaller one has radius $\overline{CA}$ and the larger one has a radius $\overline{CB}$. Notice that the two additional radii drawn into the smaller circle are equal in length to radius $\overline{CA}$.

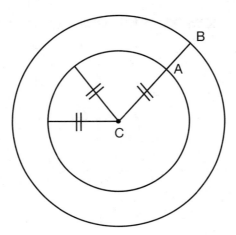

Chords and Diameters

A **chord** is a line segment that joins two points on a circle. On the following page, $\overline{AC}$ is a chord.

A **diameter** is a chord that joins two points on a circle and passes through the center point. $\overline{DB}$ is a diameter of ⊙O.

Note: A diameter is twice the length of a radius, and a radius is half the length of a diameter.

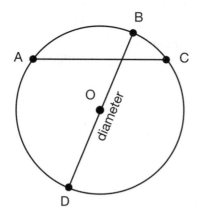

$\overline{OB}$ and $\overline{OD}$ are each a radius of ⊙O.

$\overline{DB}$ is a diameter
$\overline{AC}$ is a chord

$\overline{OB} > \overline{OD}$
$2 \times \overline{OB} = \overline{DB}$

Degrees and Arcs in Circle

Like polygons, circles contain degrees. A complete rotation around a circle contains 360°. A **semicircle** is half of a circle; it contains 180°. Any two radii in a circle form a **central angle** that will measure between 0° and 360°. In the figure below ∠AOB is a central angle.

Two radii will also form an **arc** on the outside of the circle. Arcs formed by central angles always have the same measurement as the central angle. Arc AB in the circle pictured is written $\overset{\frown}{AB}$.

A **minor arc** is any arc that measures less than 180°. $\overset{\frown}{AB}$ is a minor arc measuring 33°. A **major arc** is any arc that measures more than 180°. To name a major arc, you must include an additional letter existing on the perimeter of the circle, that is not a part of the minor arc. For example $\overset{\frown}{ADB}$ or $\overset{\frown}{ACB}$ are both ways to name the major arc measuring 327°.

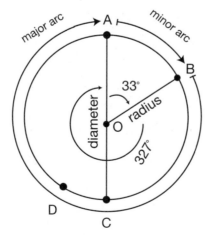

∠AOB = 33°
$\overset{\frown}{AB}$ = 33°
∠AOB = $\overset{\frown}{AB}$
$\overset{\frown}{AB}$ is a minor arc
$\overset{\frown}{ABC}$ is a semicircle
$\overset{\frown}{ABD}$ is a major arc

Congruent Arcs and Circles

Congruent circles have congruent radii and diameters. Congruent central angles form congruent arcs in congruent circles.

Tangent and Secant Lines in Circles

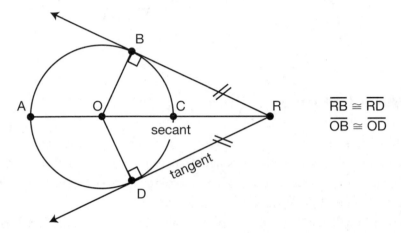

$$\overline{RB} \cong \overline{RD}$$
$$\overline{OB} \cong \overline{OD}$$

A **tangent** is a ray or line segment that intercepts a circle at exactly one point. The angle formed by a radius and a tangent where it meets a circle is **always right angle**.

Note: Two tangents from the same exterior point are congruent.

A **secant** is a ray or line segment that intercepts a circle at two points.

Set 76

Choose the best answer.

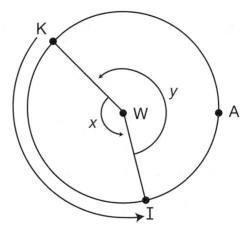

352. Which statement accurately describes what is illustrated in circle W?

    a. ∠KWI forms major arc $\overset{\frown}{KI}$

    b. ∠y forms minor arc $\overset{\frown}{KI}$

    c. ∠x forms major arc $\overset{\frown}{KI}$

    d. ∠y forms major arc $\overset{\frown}{KAI}$

353. In a circle, a radius

    a. is the same length of a radius in a congruent circle.

    b. extends outside the circle.

    c. is twice the length of a diameter.

    d. determines an arc.

354. Congruent circles

    a. have the same center point.

    b. have diameters of the same length.

    c. have radii of the same length.

    d. **b** and **c**

Use the figure below to answer question 355.

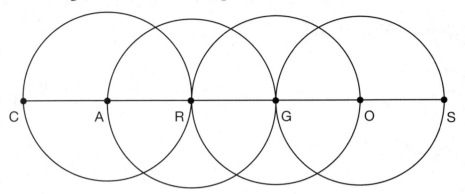

355. If the diameter of circle A is 9 cm, what is the length of $\overline{CS}$?

    a. 27 cm

    b. 4.5 cm

    c. 22.5 cm

    d. 18 cm

356. •A lies 12 inches from the center of ⊙P. If ⊙P has a 1-foot radius. •A lies

    a. inside the circle.

    b. on the circle.

    c. outside the circle.

    d. between concentric circles.

357. A diameter is also

    a. a radius.

    b. an arc.

    c. a chord.

    d. a line.

358. Both tangents and radii
    a. extend from the center of a circle.
    b. are half a circle's length.
    c. meet a circle at exactly one point.
    d. are straight angles.

359. From a stationary point, Billy throws four balls in four directions. Where each ball lands determines the endpoint for a radius of a new circle. What do the four circles have in common?
    a. a center point
    b. a radius
    c. a diameter
    d. a tangent

360. Fernando has three goats, Izzy, Eli, and Pat. He keeps them in a field, each on a leash that is tethered to a stake in the ground in a straight line, so that each goat has its own circular grazing area for the day. Izzy's leash is 40 feet long, Eli's leash is 30 feet long, and Pat is just a few months old, with a 20-foot leash. If Fernando doesn't want any the goats to be able to get any closer than 15 feet to each other, what is the minimum amount of horizontal length of field needed for these three goats?
    a. 210 feet
    b. 120 feet
    c. 135 feet
    d. 225 feet

Set 77

Use the figure below to answer question 361.

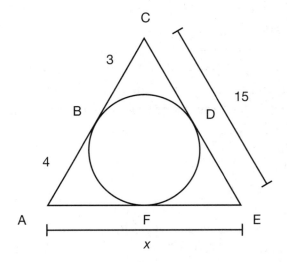

361. The above figure is *not* drawn to scale. Use your knowledge of tangents to find the value of *x*.

Use the figure below to answer question 362.

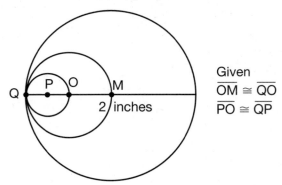

362. If the diameter of ⊙M is 2 inches, then what is the radius of ⊙P?

Use the figure below to answer question 363.

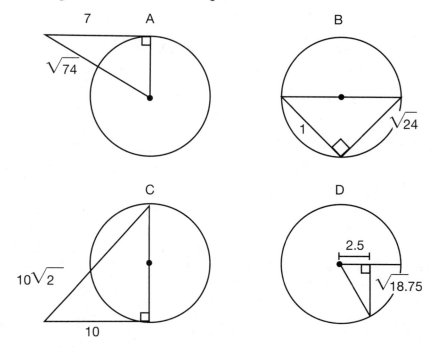

363. Which circle is NOT congruent to the other circles?

Use the figures below to answer question 364.

L. $\overset{\frown}{AB} \cong \overset{\frown}{CD}$

P. $\overset{\frown}{AB} \cong \overset{\frown}{CD}$

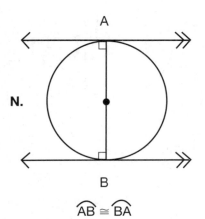

N. $\overset{\frown}{AB} \cong \overset{\frown}{BA}$

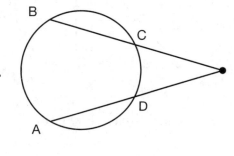

O. $\overset{\frown}{AB} \cong \overset{\frown}{CD}$

364. In which figure (L, N, P, O) is the set of arcs not congruent?

Use the figure below to answer questions 365 through 367.

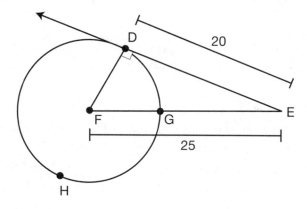

365. What is the length of a radius in the circle?

366. What is the area of ΔDEF?

367. What are the names of the minor arc and major arc in circle F?

Answers
Set 76

352. **d.** Major arcs always need a third letter to distinguish them from the minor arc that has the same endpoints. ∠y forms major arc $\overparen{KAI}$.

353. **a.** A circle is a set of points equidistant from a center point. Congruent circles have points that lie the same distance from two different center points. Consequently, the radii (the line segments that connect the center point to the points on a circle) of congruent circles are congruent. Choices **b** and **c** are incorrect because they describe secants. Choice **d** describes a chord.

354. **d.** Congruent circles have congruent radii; if their radii are congruent, then their diameters are also congruent. Choice **a** describes concentric circles, not congruent circles.

355. **c.** Since all of these circles share a radius, they are all congruent with a radius that is half of the 9 cm diameter of circle A. There are five radii, each measuring 4.5 cm, from point C to point S, which means that $\overline{CS} = 5 \times 4.5 = 22.5$ cm.

356. **b.** 12 inches is a foot, so •A lies on ⊙P. If the distance from •A to the center point measured less than the radius, then •A would rest inside ⊙P. If the distance from •A to the center point measured greater than the radius, then •A would rest outside of ⊙P.

357. **c.** A diameter is a special chord; it is a line segment that bridges a circle *and passes through the center point.*

358. **c.** As a tangent skims by a circle, it intercepts a point on that circle. A radius spans the distance between the center point of a circle and a point on the circle; like a tangent, a radius meets exactly one point on a circle.

359. **a.** Billy acts as the central fixed point of each of these four circles, and circles with a common center point are concentric.

360. **210 feet.** Each goat will have a circle with a diameter that is twice the length of its leash (which is the radius). Izzy needs 80 feet, Eli needs 60 feet, and Pat needs 40 feet. Draw these three circles so they each have 15 feet between them. The sum of their diameters and the two 15-foot gaps between them is 210 feet.

Set 77

361. $x = 16$. Don't get confused by the triangle shape that surrounds the circle. Instead, think of these legs as rays or lines that intercept the circle at one point and meet at the same exterior point. Remember from page 202 that tangent lines drawn from a single exterior point are congruent to each of their points of interception with the circle. Therefore, $\overline{AF}$ is congruent to $\overline{AB}$, $\overline{EF}$ is congruent to $\overline{ED}$, $\overline{CB}$ is congruent to $\overline{CD}$, $\overline{AB}$ is 4, and $\overline{DE}$ is the difference of $\overline{CE}$ and $\overline{CD}$, or 12; x is 4 plus 12, or 16.

362. **Radius $\odot$P = 0.25.** The diameter of $\odot$O is half the diameter of $\odot$M. Therefore, the diameter of $\odot$O is 1 in. The diameter of $\odot$P is half the diameter of $\odot$O. Therefore, the diameter of $\odot$P is 0.5 inches, making the radius 0.25.

363. **$\odot$B.** Watch out for the trap! Choices **a** and **d** show you circles with diameters. Choices **b** and **c** show you a radius. Use the Pythagorean theorem to find the length of each circle's radius:

$\odot$A: $7^2 + b^2 = (\sqrt{74})^2$. $49 + b^2 = 74$. $b^2 = 25$. $b = 5$. $Radius = 5$.

$\odot$B: $1^2 + (\sqrt{24})^2 = c^2$. $1 + 24 = c^2$. $25 = c^2$. $5 = c$. $Radius = \frac{1}{2}(5) = 2.5$.

$\odot$C: $10^2 + b^2 = (10\sqrt{2})^2$. $100 + b^2 = 200$. $b^2 = 100$. $b = 10$. $Radius = \frac{1}{2}(10) = 5$.

$\odot$D: $2.5^2 + (\sqrt{18.75})^2 = c^2$. $6.25 + 18.75 = c^2$. $25 = c^2$. $5 = c$. $Radius = 5$.

Only $\odot$B is not congruent to $\odot$A, $\odot$C, and $\odot$D.

364. **$\odot$O.** Parallel and congruent line segments form congruent arcs in $\odot$L. Two diameters form equal central angles which form congruent arcs in $\odot$P. Parallel tangent lines form congruent semi-circles in $\odot$N. Secants extending from a fixed exterior point form non-congruent arcs in $\odot$O.

365. *Radius* = **15.** The angle formed by a radius and a tangent where it meets the circle is always a right angle. Use the Pythagorean theorem to find the length of the radius, $\overline{DF}$: $a^2 + 20^2 = 25^2$. $a^2 + 400 = 625$. $a^2 = 225$. $a = 15$.

366. *Area* = **150 in³.** The length of $\overline{ED}$ is the height of ΔDEF. To find the area of ΔDEF, plug the measurements of the radius and the height into $\frac{1}{2}bh$: $\frac{1}{2}(15 \text{ in.} \times 20 \text{ in.}) = \frac{1}{2}(300 \text{ in.}^2) = 150$ square inches.

367. $\overparen{\text{DG}}$ **and** $\overparen{\text{DHG}}$**.** The minor arc is $\overparen{\text{DG}}$. Major arcs always need a third letter to distinguish them from the minor arc that has the same endpoints, so the major arc is $\overparen{\text{DHG}}$.

17

Circumference and Area of Circles

The Importance of π in Circles

The irrational number π (which is spelled "pi" and pronounced "pie"), is a key concept when working with circles. *Pi* is essential to calculating the area of circles as well as the **circumference**—the perimeter around the outside—of a circle. Over 2,000 years ago Greek mathematicians discovered that π is actually the never-changing quotient of the $\frac{circumference}{diameter}$. This ratio is the same in all circles, and although *pi* is an irrational number that continues infinitely, the most commonly used approximation for π is 3.14.

The Circumference of a Circle

Remember that circumference is just the name for perimeter that is used with circles. You can use the diameter or the radius to find the circumference of a circle by using either of these formulas:

Circumference = πd or $2\pi r$

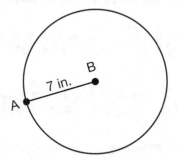

C = $2\pi r$
C = $2 \times \pi \times 7$ inches
C = 14π inches

Using Circumference to Measure Arcs

Since you now know how to find the entire circumference of a circle, you can apply that to measuring arc lengths made by central angles. This is illustrated below.

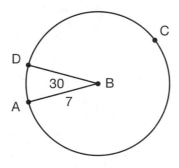

$C = 14\pi$ inches

$$\frac{\text{part}}{\text{whole circle}} = \frac{30°}{360°} = \frac{1}{12}$$

$\overset{\frown}{AD}$ is $\dfrac{1}{12}$ of the 14π circumference

So $\overset{\frown}{AC} = \dfrac{1}{12} \times 14\pi = \dfrac{7}{6}\pi$

Area of a Circle

Pi is also essential when calculating the area of a circle. Unlike with circumference, only the *radius* can be used to calculate area. This means if you are given the diameter of a circle, you will need to divide it by two to get the radius before finding the area. **Area = πr^2**

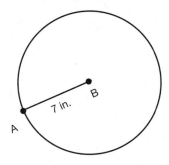

$A = \pi r^2$

$A = \pi(7 \text{ inches})^2$

$A = 49\pi$ square inches

Set 78

Choose the best answer.

368. What is the circumference of the figure below?

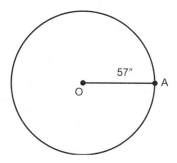

 a. 57π inches
 b. 114π inches
 c. 26.5π inches
 d. √57π inches

369. What is the area of the figure below?

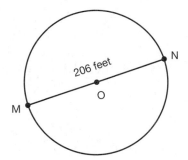

 a. 51.5π square feet
 b. 103π square feet
 c. 206π square feet
 d. 10,609π square feet

370. If $\overline{ME}$ = 10x, what expression represents the area of Circle A?

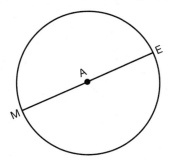

    a. $25x^2\pi$
    b. $20x^2\pi$
    c. $100x^2\pi$
    d. $100x\pi$

371. The area of a square is 484 square feet. What is the maximum area of a circle inscribed in the square (where the diameter of the circle = the length of one square side)?
    a. 11π square feet
    b. 22π square feet
    c. 484π square feet
    d. 122π square feet

372. If the circumference of a circle is 192π feet, then the length of the circle's radius is
    a. $16\sqrt{6}$ feet.
    b. 96 feet.
    c. 192 feet.
    d. 384 feet.

373. If the area of a circle is 289π square feet, then the length of the circle's radius is
    a. 17 feet.
    b. 34 feet.
    c. 144.5 feet.
    d. 289 feet.

374. What is the area of a circle inscribed in a dodecagon with an apothem 13 meters long?
- **a.** 26π square meters
- **b.** 156π square meters
- **c.** 42.2π square meters
- **d.** 169π square meters

Use the figure below to answer questions 375 through 376. It is NOT drawn to scale

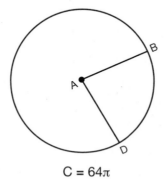

C = 64π

375. $\overset{\frown}{BD}$ is a quarter of the circumference of ⊙A. If the total circumference of ⊙A is 64π feet, then what is the length of $\overset{\frown}{BD}$?
- **a.** 16π feet
- **b.** 32π feet
- **c.** 48π feet
- **d.** 90π feet

376. What is the central angle, ∠BAD that intercepts $\overset{\frown}{BD}$?
- **a.** an acute angle
- **b.** a right angle
- **c.** an obtuse angle
- **d.** a straight angle

Use the figure below to answer question 377.

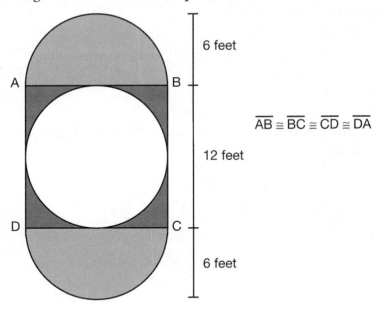

$\overline{AB} \cong \overline{BC} \cong \overline{CD} \cong \overline{DA}$

377. What is the area of all of the shaded regions in the figure.
   a. 144 square feet – 12π square feet
   b. 12 square feet – 144π square feet
   c. 144 square feet
   d. 144 square feet – 24π square feet + 12π square feet

Set 79

Use the figure below to answer questions 378 through 379.

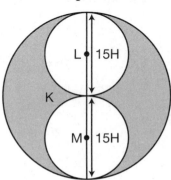

378. What is the area of the shaded regions in the figure?
   a. 56.25π square feet
   b. 112.5π square feet
   c. 225π square feet
   d. 337.4π square feet

379. What is the ratio of the area of ⊙M and the area of ⊙K?
   a. 1:8
   b. 1:4
   c. 1:2
   d. 1:1

Use the figure below to answer questions 380 through 381.

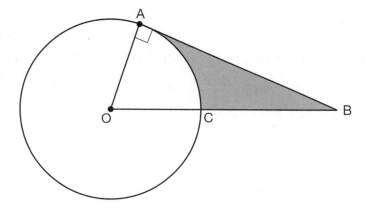

380. If $\overline{AB}$ = 60 and $\overline{OB}$ = 75, what is the area of ⊙O?

381. If ∠AOC measures 60°, what is the area of the slice contained within the circle, between radii $\overline{OA}$ and $\overline{OC}$?

382. Continuing with the information that ∠AOC measures 60°, what is the area of the shaded region of the figure?

Using the figure below answer questions 383 through 387.

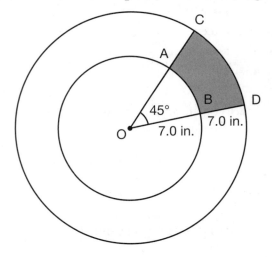

383. An outer ring exists outside of the circle that has a radius of $\overline{OA}$, but inside the circle that has a radius of $\overline{OC}$. What is the area of this outer ring?

384. Find the shaded area of the figure.

385. Find the length of $\overset{\frown}{AB}$.

386. Find the length of $\overset{\frown}{CD}$.

387. Although they are defined by the same angle, what is the relationship between the lengths of arcs $\overset{\frown}{CD}$ and $\overset{\frown}{AB}$? What does this tell you about how arc length relates to angle size and radius length?

Set 80

Use the figure below to answer questions 388 through 389.

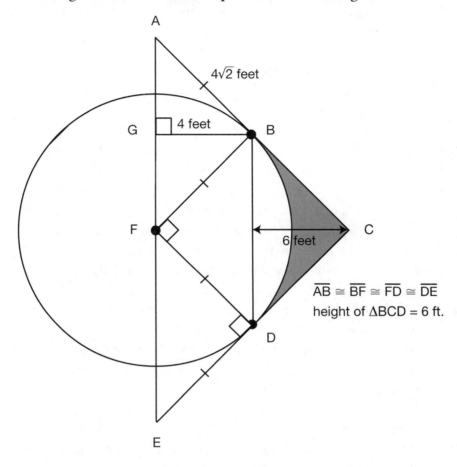

A

$4\sqrt{2}$ feet

G ⌐ 4 feet B

F ⌐ C

6 feet

$\overline{AB} \cong \overline{BF} \cong \overline{FD} \cong \overline{DE}$
height of $\triangle BCD$ = 6 ft.

D

E

388. What is the area of trapezoid ABDE?

389. What is the area of the shaded region?

Answers
Set 78

368. **b.** The perimeter of a circle is twice the radius times *pi*: $(2 \times 57$ inches$)\pi$.

369. **d.** You first need to divide the given diameter of 206 feet, in two in order to find the radius. The area of a circle is the radius squared times *pi*: $\pi(103 \text{ feet})^2$.

370. **a.** Follow the same first step as you did in question 369. Since $\overline{ME} = 10x$, the radius is $5x$. Area $= \pi(5x)^2 = \pi(25x^2) = 25x^2\pi$

371. **c.** If the area of a square is 484 square feet, then the sides of the square must measure 22 feet each. The diameter of an inscribed circle has the same length as one side of the square. The maximum area of an inscribed circle is $\pi(11 \text{ feet})^2$, or 121π square feet.

372. **b.** Backsolve by plugging the given circumference, 192π feet, into the formula for circumference of a circle. The circumference of a circle is *pi* times twice the radius. 192 feet is twice the length of the radius; therefore half of 192 feet, or 96 feet, is the actual length of the radius.

373. **a.** Just as in question 372, backsolve by plugging 289π into the formula for area. The area of a circle is *pi* times the square of its radius. If 289 feet is the square of the circle's radius, then 17 feet is the length of its radius. Choice **c** is not the answer because 144.5 is half of 289, not the square root of 289.

374. **d.** Have you forgotten these terms? Look back to chapter 13 on polygons for a refresher. In the mean time, don't get too distracted by the shape of the polygon, in this case a dodecagon. Hold fast to two facts—the circle is set within the dodecagon and the apothem goes to the dodecagon's center. Therefore, the center of the circle overlaps with the center of the dodecagon and, accordingly, the apothem = the radius. From here, solve for area. $A = \pi r^2$. $A = \pi(13)^2$. $A = 169\pi$ square meters.

375. **a.** The length of arc BD is a quarter of the circumference of ⊙A, or 16π feet.

376. **b.** A quarter of 360° is 90°; ∠BAD is a right angle.

377. **c.** This question is much simpler than it seems. The half circles that cap square ABCD form the same area as the circular void in the center. Find the area of square ABCD, and that is your answer. 12 feet × 12 feet = 144 feet. Choice **a** and **d** are the same answer. Choice **b** is a negative area and is incorrect.

Set 79

378. **b.** The radii of ⊙L and ⊙M are half the radius of ⊙K. Their areas equal π(7.5 feet)^2, or 56.25π square feet each. The area of ⊙K is π(15^2), or 225π square feet. Subtract the areas of circles L and M from the area of ⊙K: 225π sq. ft. – 112.5π sq. ft. = 112.5π square feet.

379. **b.** A ratio is a comparison of two quantities, as discussed in chapter 8. You'll need to compare the values for area you found in question 378. Though ⊙M has half the radius of ⊙K, it has a fourth of the area of ⊙K. 56.25π square feet: 225.0π square feet, or 1:4.

380. **Area = 2,025 π square feet.** You will need to use your mastery of the Pythagorean theorem to solve for leg $\overline{OA}$. Then plug that value into the area for a circle, as $\overline{OA}$ is also the radius: $a^2 + 60^2 = 75^2$. $a^2 + 3,600 = 5,625$. $a^2 = 2,025$. $a = 45$. Then put this radius into A = πr^2, A = π(45)^2 = 2,025π square feet.

381. **337.5π.** Since the area of circle O is 2,025π square feet and $m\angle AOC = 60°$, set up a proportion to calculate the area of the slice of circle contained by radii $\overline{OA}$ and $\overline{OC}$. Do this by comparing the part to whole ratio of the angle measurements to the part to whole ratio of the areas:

$\frac{part}{whole} = \frac{60°}{360°} = \frac{\text{partial area}}{2025\pi}$

360 (partial area) = 60(2025π)

partial area = $\frac{60(2025\pi)}{360}$ = 337.5π

382. **1,350 – 337.5π square units.** The area of ΔABO is A = $(\frac{1}{2})(b)(h)$ = $(\frac{1}{2})(60)(45)$ = 1,350. Subtract the area of the slice of circle found in question 381 from 1,350 to get the area of the shaded region: 1,350 – 337.5π square units.

383. **147π in².** The area of the larger circle with the radius of 14 is π(14)² or 196π in². The area of the smaller circle with the radius of 7 is π(7)² or 49π in². Subtract these two areas to find the area of the outer ring: 196π – 49π = 147π in².

384. **18.4π in².** As found in the previous question, the area of the outer ring is 147π in². Since the shaded region is defined by a 45° angle and $\frac{45°}{360°}$ = $\frac{1}{8}$, multiply 147π by $\frac{1}{8}$ to get 18.4π in².

385. **1.75π inches.** The circumference of small ⊙O is 14π inches. A 45° slice of that circumference is one-eighth the circumference, or 1.75π inches.

386. **3.5π inches.** The circumference of concentric ⊙O is 28π inches. An eighth of that circumference is 3.5π inches.

387. **Arc $\overarc{CD}$ is twice as long as arc $\overarc{AB}$.**
This shows that just because two arcs are defined by the same angle, they will not necessarily have the same arc length. If the angle measurement is the same, the bigger the radius of the circle, the bigger the arc length.

Set 80

388. *Area* **= 48 square feet.** Use the Pythagorean theorem to find $\overline{AG}$. $(4\sqrt{2})^2 = (4)^2 + b^2$. $32 = 16 + b^2$. $b = 4$. If $\overline{AG}$ equals 4 feet, then $\overline{AF}$, $\overline{EF}$ and $\overline{BD}$ equal 8 feet, and $\overline{AE}$ equals 16 feet. The area of a trapezoid is half its height times the sum $\frac{1}{2}$of its bases: $\frac{1}{2}$(4 ft.)(8 ft. + 16 ft.) = 2(24) = 48 square feet.

389. *Area* ≈ **14.88 square feet.** The shaded area is the difference of ΔBCD's area and the area between chord BD and arc BD. The height of ΔBCD is 6 feet. Its area is $\frac{1}{2}(6 \times 8) = 24$ sq. ft. The area of $\overset{\frown}{BD}$ is tricky. It is the area of the circle contained within ∠BFD minus the area of inscribed ΔBFD. Central angle BFD is a right angle; it is a quarter of a circle's rotation and a quarter of its area. The circle's radius is $4\sqrt{2}$ feet. The area of circle F is $\pi(4\sqrt{2}$ ft.$)^2$, or 32π square feet. A quarter of that area is 8π square feet. The area of ΔBFD is $\frac{1}{2}(4\sqrt{2} \times 4\sqrt{2}) = 16$ sq. ft. Subtract 16 square feet from 8π square feet, then subtract that answer from 24 square feet and your answer is approximately 14.88 square feet.

18

Working with Cylinders, Cones, and Spheres

Up until this point, you have only learned how to calculate surface area and volume of prisms that have flat sides and edges. Now that you are familiar with π, you are ready to learn how to work with cylinders, cones, and spheres.

Cylinders

The most common **cylinder** that you use every day might be a drinking glass or a soda can. Cylinders are made up of two round bases connected by a singular tube of parallel sides. The length of the "tube" part of the cylinder is considered its height, which is critical in order to calculate the surface area and volume of a cylinder:

The surface area of a cylinder is: $SA = 2\pi r^2 + (2\pi r)(h)$

Think of an empty paper towel roll cut lengthwise and rolled out to form a rectangle. Calculating the surface area of a cylinder consists of adding the two areas for the circle bases ($2\pi r^2$) to the area of the rectangle "tube." The width of that tube is actually the circumference of the circular base ($2\pi r$).

Therefore, the $(2\pi r)(h)$ in the formula is actually a calculation of the side of the cylinder.

The volume of a Cylinder is: $V = (\pi r^2)(h)$

Calculating the volume of a cylinder is similar to calculating the volume of a right prism, in that it is the (area of the base)(height). This is an easy way to remember the formula!

Cones

A cone is similar to a cylinder, except it only has a circular base on one end, and the other end is a single point. The distance from that singular point to the center of the circular base is the height of the cone. The volume of a cone is $\frac{1}{3}$ of the volume of a cylinder with the same height and radius:

The volume of a Cone is: $V = \frac{1}{3}(\pi r^2)(h)$

Spheres

A sphere is a set of points equidistant from one central point. This is the mathematical definition for the shape of a basketball! Sphere's are only measured by the length of their radii; they do not have a separate height like cylinders and cones.

The surface area of a Sphere is: $SA = 4\pi r^2$

The surface area of a sphere is four times the area of a circle.

The volume of a Sphere is: $V = \frac{4}{3}\pi r^3$

Set 81

Use the figure below to answer questions 390 through 392.

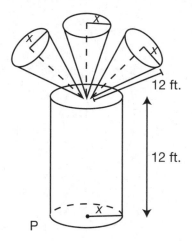

Volume of cylinder P = 432π cubic ft.

390. If the volume of the cylinder P is 432π cubic feet, what is the length of x?

391. What is the surface area of cylinder P?

392. What is the total combined volume of the cylinder and the three pictured cones?

Set 82

Use the figure below to answer questions 393 through 395.

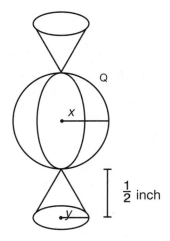

Volume Q = $\frac{1}{6}$ cubic inches

393. If the volume of the spherical part of candy wrapper Q is $\frac{1}{6}\pi$ cubic inches, what is the length of x?

394. If the conical ends of candy wrapper Q have $\frac{1}{96}\pi$ cubic inch volumes each, what is the length of y?

395. What is the surface area of the gumball inside the wrapper?

Set 83

Solve each question using the information in each word problem.

396. Tracy and Jarret try to share an ice cream cone, but Tracy wants half of the scoop of ice cream on top while Jarret wants the ice cream inside the cone. Assuming the half scoop of ice cream on top is part of a perfect sphere, who will have more ice cream? The cone and scoop both have radii 1 inch long; the cone is 3 inches high. Assume the cone is filled with ice cream.

397. Dillon fills the cylindrical coffee grind containers each day. One bag has 32π cubic inches of grinds. How many cylindrical containers can Dillon fill with two bags of grinds if each cylinder is 4 inches wide and 4 inches high?

398. Jaap is making all of the table arrangements for Christine's wedding. She has chosen a design that places an 18-inch tall cylindrical vase with a diameter of 6 inches into another 18-inch tall cylindrical vase that has an eight-inch diameter. Sea grasses and flowers will be put into the inner vase and the gap between the two vases will be filled with sand. If Christine will have 12 tables at her wedding, how many cubic inches of sand does Jaap need to purchase to complete this order? (Keep your answer in π form.)

399. A farm in the Midwest has a grain silo that is a round cylinder with half of a dome on top. The height of the silo from the ground to the top of the dome is 22 meters. The width of the bottom of the silo is six meters. What is the total volume of the silo? (Keep your answer in π form.)

400. Munine is trying to fit her new 24-inch tall cylindrical speakers in a corner shelf. Unfortunately, they do not fit upright in her shelf. If each speaker is $2,400\pi$ cubic inches, what is the maximum width of her shelf.

401. Tory knows that the space in a local cathedral dome is $13,122\pi$ cubic feet. Using her knowledge of geometry, what does Tory calculate the height of the dome to be?

Set 84

402. In art class, Billy adheres 32 identical half spheres to cover with canvas. What is their total surface area, not including the flat side adhered to the canvas, if the radius of one sphere is 8 centimeters?

403. Joe carves a perfect 3.0-meter wide sphere inside a right prism. If the volume of the prism is 250.0 cubic meters, how much material did he remove? How much material remains?

404. Theoretically, how many spherical shaped candies should fit into a cylindrical jar if the diameter of each candy is 0.50 inch, and the jar is 4.50 inches wide and 6 inches long?

405. A sphere with a 2-foot radius rests inside a cube with edges 4.5 feet long. What is the volume of the space between the sphere and the cube assuming *pi* ≈ 3.14?

Set 85

Use Puppet Dan to answer questions 406 through 414.

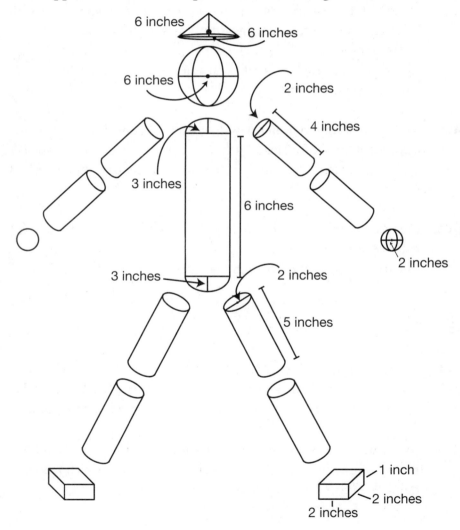

6 inches   6 inches

6 inches

2 inches

4 inches

3 inches

6 inches

2 inches

3 inches

2 inches

5 inches

1 inch

2 inches

2 inches

406. What is the volume of Puppet Dan's hat if it measures 6 inches wide by 6 inches high?

407. What is the volume of Puppet Dan's head if it measures 6 inches wide?

408. What is the volume of Puppet Dan's arms if one segment measures 2 inches wide by 4 inches long?

409. What is the volume of Puppet Dan's hands if each one measures 2 inches wide?

410. What is the volume of Puppet Dan's body if it measures 6 inches wide and 6 inches long? Each end of the cylinder measures 6 inches wide.

411. What is the volume of Puppet Dan's legs if each segment measures 2 inches wide by 5 inches long?

412. What is the volume of Puppet Dan's feet if each foot measures 2 inches × 2 inches × 1 inch?

413. What is puppet Dan's total volume?

414. Puppet Dan is made out of foam. If foam weighs 3 ounces per cubic inch, how much does the total of puppet Dan's parts weigh?

Answers

Set 81

390. *x* = **6 feet.** The radius of cylinder P is represented by x; it is the only missing variable in the volume formula. Plug it in and solve for x: 432π cubic ft. = $(\pi x^2)12$ ft. 36 sq. ft. = x^2. 6 feet = x.

391. *Surface area* = **216π square feet.** The surface area of a cylinder is $2\pi r^2 + 2\pi rh$: Plug the variables in and solve: $SA = 2\pi(6\text{ ft})^2 + 2\pi(6$ ft. × 12 ft.). 72π sq. ft.+ 144π sq. ft. = 216π sq. ft.

392. *Total volume* = **864π cubic feet.** This problem is easier than you think. Each cone has exactly the same volume. The three cones together equal the volume of the cylinder because a cone has $\frac{1}{3}$ the volume of a cylinder with the same height and radius. Multiply the volume of the cylinder by 2, and you have the combined volume of all three cones and the cylinder.

Set 82

393. *x* = $\frac{1}{2}$ **inch.** The formula for volume of a sphere is πr^3. In this question x is the value of r. Plug the variables in and solve for x: $\frac{1}{6}\pi = \pi x^3$. $\frac{1}{6} = x^3$. $\frac{1}{2} = x$.

394. *y* = $\frac{1}{4}$ **inch.** The volume of a cone is $\frac{1}{3}\pi r^2 h$, where y is the value of r. Plug in the variables and solve: $\frac{1}{96}\pi$ cubic in. = $\frac{1}{3}\pi y^2 2\frac{1}{2}$ in. $\frac{1}{96}\pi$ cubic in. = $\frac{1}{6}\pi y^2$. $\frac{1}{16}\pi$ sq. in. = y^2. $\frac{1}{4}$ inch = y.

395. *Surface area* = **1.0π square inch.** The candy inside the wrapper is a perfect sphere. The formula for its surface area is $4\pi r^2$. Plug the variables in and solve: $SA = 4\pi(0.5\text{ inch})^2$. $SA = 1.0\pi$ square inch.

Set 83

396. **Jarret.** The volume of a half sphere is $\frac{1}{2}(\frac{4}{3}\pi r^3)$. Tracy's half scoop is then $\frac{1}{2}(\frac{4}{3}\pi \times 1 \text{ inch}^3)$, or $\frac{2}{3}\pi$ cubic inches. The volume of a cone is $\frac{1}{3}\pi r^2 h$. The ice cream in the cone is $\frac{1}{3}\pi(1 \text{ inch}^2 \times 3 \text{ inches})$, or π cubic inches. Jarret has $\frac{1}{3}\pi$ cubic inches more ice cream than Tracy.

397. **4 containers.** Remember that the given measurement of 4 inches wide is equal to the diameter of the cylinder. You need to take half this measurement in order to determine the radius. The volume of each container is $\pi(2 \text{ in.})^2(4 \text{ in.})$, or 16π cubic inches. One bag fills the volume of two containers. Two bags will fill the volume of four containers.

398. **$1,512\pi$ in³.** First find the volume of the larger vase, using a radius of 4: $V = (\pi 4^2)(18) = 288\pi$ in³. Next find the volume of the smaller vase, using a radius of 3: $V = (\pi 3^2)(18) = 162\pi$ in³. Subtract these two volumes to find the volume of the gap between the vases: $288\pi - 162\pi = 126\pi$ in³. This is the volume of sand needed for each vase, so multiplying that by 12 we get $126\pi \times 12 = 1,512\pi$ in³.

399. **189π m².** Since the base of the cylinder is six meters wide, it has a radius of three meters. Therefore the dome will extend 3 meters above the top of the cylindrical base, making height of the cylindrical base 19 meters. Calculate the volume of a cylinder that is 19 meters tall with a radius of three meters: $V = (\pi 3^2)(19) = 171\pi$ m². Next find half of the volume of the sphere-shaped dome on top that has a radius of 3: $\frac{1}{2}(\frac{4}{3}\pi 3^3) = 18\pi$ m². Add these together to find the total volume: 171π m² $+ 18\pi$ m² $= 189\pi$ m².

400. **Less than 20 inches.** The volume of a single speaker is $\pi(r^2 \times 24$ inches$) = 2,400\pi$ cubic inches. Now solve for the radius. $r^2 = 100$ square inches. $r = 10$ inches. The width of each speaker is twice the radius, or 20 inches. Munine's shelf is less than 20 inches wide.

401. **27 feet.** Half the volume of a sphere is $\frac{1}{2}(\frac{4}{3}\pi r^3)$, or $\frac{2}{3}\pi r^3$. If the volume is $13,122\pi$ cubic feet, then the radius is 27 feet ($\sqrt[3]{19,683} = 27$). The height of the dome is equal to the radius of the dome; therefore the height is also 27 feet.

Set 84

402. **$4,096\pi$ square centimeters.** Surface area of a whole sphere is $4\pi r^2$. The surface area of half a sphere is $2\pi r^2$. Each sphere's surface area is $2\pi(8 \text{ centimeters}^2)$, or 128π square centimeters. Now, multiply the surface area of one half sphere by the 32 halves: $32 \times 128\pi$ square centimeters $= 4,096\pi$ square centimeters.

403. **Approximately 235.9 cubic meters.** Joe removed the same amount of material as volume in the sphere, or $\frac{4}{3}\pi(1.5 \text{ meters})^3$, which simplifies to 4.5π cubic meters. The remaining volume is 250 cubic meters – 4.5π cubic meters, or approximately 235.9 cubic meters.

404. **1,518 candies.** The volume of each piece of candy is $\frac{4}{3}\pi(0.25 \text{ inches})^3$, or 0.02π cubic inches. The volume of the jar is $\pi(2.25 \text{ inches}^2 \times 6)$ inches, or 30.375π cubic inches. Divide the volume of the jar by the volume of a candy $\frac{30.375\pi \text{ cubic inches}}{0.02\pi \text{ cubic inches}}$, and 1,518 candies can theoretically fit into the given jar (not including the space between candies).

405. *Remaining volume* $\approx$ **57.6 ft.** First, find the volume of the cube, which is $(4.5 \text{ feet})^3$, or approximately 91.1 cubic feet. The volume of the sphere within is only $\frac{4}{3}\pi(2 \text{ feet})^3$, or approximately 33.5 cubic feet. Subtract the volume of the sphere from the volume of the cube. The remaining volume is approximately 57.6 cubic feet.

Set 85

406. $V = 18\pi$ **cubic inches.** Volume of a cone = $\frac{1}{3}\pi r^2 h$. $V = \frac{1}{3}\pi(3 \text{ in.})^2(6 \text{ in.})$.

407. $V = 36\pi$ **cubic inches.** Volume of a sphere = $\frac{4}{3}\pi r^3$. $V = \frac{4}{3}\pi(3 \text{ in.})^3$.

408. 16π **cubic inches.** Volume of a cylinder = $\pi r^2 h$. $V = \pi(1 \text{ in.}^2 \times 4 \text{ in.})$ $V = 4\pi$ cubic inches. There are four arm segments, so four times the volume = 16π cubic inches.

409. $\frac{8}{3}\pi$ **cubic inches.** Volume of a sphere = $\frac{4}{3}\pi r^3$. $V = \frac{4}{3}\pi(1 \text{ in.}^3)$. $V = \frac{4}{3}\pi$ cubic inches. There are two handballs, so two times the volume = $\frac{8}{3}\pi$ cubic inches.

410. 90π **cubic inches.** The body is the sum of two congruent half spheres, which is really one sphere, and a cylinder. *Volume of a sphere* = $\frac{4}{3}\pi r^3$. $V = \frac{4}{3}\pi(3 \text{ in.})^3$. $V = 36\pi$ cubic inches. *Volume of a cylinder* = $\pi r^2 h$. $V = \pi(3 \text{ in.})^2 (6 \text{ in.})$; $V = 54\pi$ cubic inches. *Total volume* = 90π cubic inches.

411. **20π cubic inches.** Volume of a cylinder = $\pi r^2 h$. $V = \pi(1 \text{ in.}^2 \times 5 \text{ in.})$ $V = 5\pi$ cubic inches. There are four leg segments, so four times the volume = 20π cubic inches.

412. **8 cubic inches.** Each foot is a rectangular prism. *Volume of a prism = length × width × height.* $V = 2 \text{ in.} \times 2 \text{ in.} \times 1 \text{ in.}$ $V = 4$ cubic inches. There are two feet, so two times the volume = 8 cubic inches.

413. $V \approx$ **581.36 cubic inches.** The sum of the volumes of its parts equals a total volume. 18π cubic inches + 36π cubic inches + 16π cubic inches + $\frac{8}{3}\pi$ cubic inches + 90π cubic inches + 20π cubic inches ≈ 182.6π cubic inches + 8 cubic inches. If π ≈ 3.14, then $V \approx 581.36$ cubic inches.

414. **1,744.08 ounces.** Multiply: $\frac{3 \text{ ounces}}{1 \text{ cubic inch}} \times 581.36$ cubic inches = 1,744.08 ounces. Puppet Dan is surprisingly light for all his volume!

19

Coordinate Geometry

In this chapter we will learn how to graph points on a coordinate plane and find the distance between any two points. The **coordinate plane** is a special type of map used in mathematics to plot points, lines, and shapes. The coordinate plane is made up of two main axes that intersect at **the origin**, to form a right angle. The *x*-axis is the main horizontal axis, and the *y*-axis is the main vertical axis. These perpendicular axes divide the coordinate plane into four sections, or **quadrants**. The quadrants are numbered in counter-clockwise order, beginning with the top right quadrant.

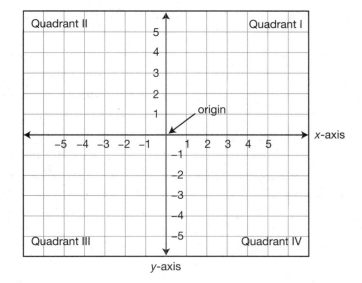

Plotting Points on a Coordinate Plane

Points in the coordinate plane have an x-coordinate and a y-coordinate that show where the point is in relation to the origin. Points are presented in the format (x, y).

   A point's position left or right of the origin is its x-coordinate; a point's position up or down from the x-axis is its y-coordinate. Every point's coordinate pair shows the number of spaces left or right of the y-axis and the number of spaces above or below the x-axis.

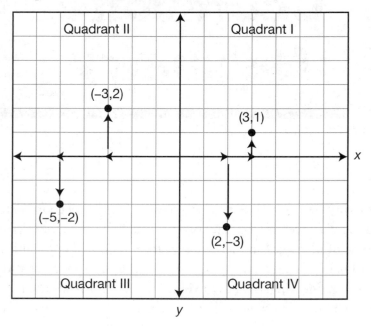

Plotting a Point on a Coordinate Plane

To plot a point you always start from the origin, which is the point $(0,0)$. Using the first coordinate, count the number of spaces indicated right ($x >$ 0) or left ($x < 0$) of the origin. Using the second coordinate, count the number of spaces indicated up ($y > 0$) or down ($y < 0$) of the x-axis.

   For example, the point $(3,1)$ in Quadrant I was plotted by starting at the origin, moving three spaces to the right, and then one space up. Similarly, the point $(-5, -2)$ in Quadrant III was plotted by starting at the origin, moving five spaces to the left (because the x-coordinate is number), and then two spaces down (because the y-coordinate is also negative).

The Length of a Line

The distance between two points on a coordinate plane is the shortest route between the two points. Unless the two points have the same x- or y-coordinate, the distance between them will be represented by a diagonal line. This diagonal line is always the hypotenuse of a right triangle that exists (but is not drawn) in the coordinate plane. (One of the legs of this imaginary triangle is a horizontal line, parallel to the x-axis and the other leg is a vertical line, parallel to the y-axis.) The length of the diagonal hypotenuse is the square root of the sum of the squares of the two legs. This is the Pythagorean theorem in reverse. (Note! In the following equations, the subscript numbers are used to distinguish the points from each other: (x_1, y_1) means point one, and (x_2, y_2) means point two.)

The *distance*, *d*, between any two points $A(x_1, y_1)$ and $B(x_2, y_2)$ in the coordinate plane is $d = \sqrt{(x_2 - x_1)^2 + (y_2 - y_1)^2}$

$c^2 = a^2 + b^2$ (Pythagorean theorem)

a = horizontal leg: $(x_2 - x_1)$
b = vertical leg: $(y_2 - y_1)$
$c = d$ (the distance between two points)
$d = \sqrt{(x_2 - x_1)^2 + (y_2 - y_1)^2}$

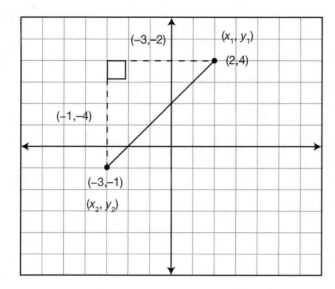

Pythagorean theorem
$a^2 + b^2 + c^2$
$\sqrt{a^2 + b^2} = c$

Distance $= \sqrt{(x_2 - x_1)^2 + (y_2 - y_1)^2}$
$D = \sqrt{(-3 - 2)^2 + (-1 - 4)^2}$
$D = \sqrt{(-5)^2 + (-5)^2}$
$D = \sqrt{25 + 25}$
$D = \sqrt{50}$
$D \approx 7.1$

Set 86

Choose the best answer.

415. The origin is
   a. where the x-axis begins.
   b. where the y-axis begins.
   c. where the x-axis intersects the y-axis.
   d. does not have a coordinate pair.

416. **True or False:** $\sqrt{(x_2 - x_1)^2 + (y_2 - y_1)^2} = \sqrt{(x_2 - x_1)^2} + \sqrt{(y_2 - y_1)^2}$

417. **True or False:** The distance between two points can always be counted by hand accurately as long as the points are plotted on graph paper.

418. •A (−3,−2) lies in quadrant
   a. I.
   b. II.
   c. III.
   d. IV.

419. •R is 3 spaces right and one space above •P (−1,−2). •R lies in quadrant
   a. I.
   b. II.
   c. III.
   d. IV.

420. •B is 40 spaces spaces left and 20 spaces below •A (20,18). •B lies in quadrant
   a. I.
   b. II.
   c. III.
   d. IV.

421. •O is 15 spaces right and 15 spaces below •N (−15,0). •O lies on
   a. x-axis.
   b. y-axis.
   c. z-axis.
   d. the origin.

422. On a coordinate plane, $y = 0$ is
   a. the x-axis.
   b. the y-axis.
   c. a solid line.
   d. finitely long.

423. A baseball field is divided into quadrants. The pitcher is the point of origin. The second baseman and the hitter lie on the y-axis; the first baseman and the third baseman lie on the x-axis. If the hitter bats a ball into the far left field, the ball lies in quadrant
   a. I.
   b. II.
   c. III.
   d. IV.

424. •A (12,3), •B (0,3) and •C (–12,3) form
  a. a vertical line
  b. a horizontal line
  c. a diagonal line
  d. a plane

425. •G (14,–2), •H (–1,15) and •I (3,0)
  a. determine a plane.
  b. are collinear.
  c. are noncoplanar.
  d. are a line.

426. The distance between •J (4,–5) and •K (–2,0) is
  a. $\sqrt{11}$.
  b. $\sqrt{29}$.
  c. $\sqrt{61}$.
  d. $\sqrt{22}$.

Set 87

State the coordinate pair for each point.

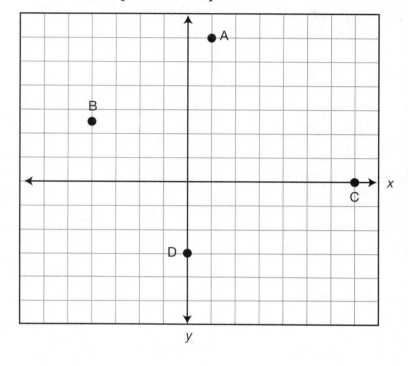

427. •A

428. •B

429. •C

430. •D

Set 88

Plot each point on the same coordinate plane. Remember to label each point appropriately.

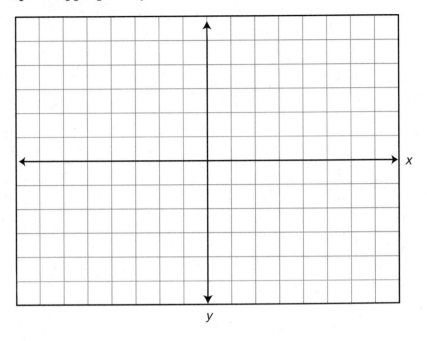

431. From the origin, plot •M (4,5).

432. From the origin, plot •N (12,–1).

433. From the origin, plot •O (–3,–6).

434. From the origin, plot •P (0,5)

435. What is the horizontal distance between •M and •O? What is the vertical distance?

436. Find the length of $\overline{MO}$.

Set 89

Find the distance between each given pair of points.

437. •A (0,4) and •B (0,32)

438. •C (−1,−2) and •D (4,−1)

439. •E (−3,3) and •F (7,3)

440. •G (17,0) and •H (−3,0)

Answers

Set 86

415. **c.** The origin, whose coordinate pair is (0,0), is in fact a location. It is where the *x*-axis meets the *y*-axis. It is not the beginning of either axis because both axes extend infinitely in opposite directions, which means they have no beginning and no end.

416. **False:** You must find the sum inside the parentheses before taking the square root of the terms.

417. **False:** Unless the coordinate pairs have the same *x*- or *y*-coordinate, the distance between them will be a diagonal line which cannot be counted by hand.

418. **c.** Both coordinates are negative: count three spaces left of the origin; then count two spaces down from the *x*-axis. •A is in quadrant III.

419. **d.** To find a new coordinate pair, add like coordinates: 3 + (–1) = 2. 1 + (–2) = –1. This new coordinate pair is •R (2,–1); •R lies in quadrant IV.

420. **c.** To find a new coordinate pair, add like coordinates: *x* will be (–40) + 20 = –20; *y* will be (–20) + 18 = –2. •B (–20, –2) lies in Quadrant III.

421. **b.** To find a new coordinate pair, add like coordinates: 15 + (–15) = 0. (–15) + 0 = –15. This new coordinate pair is (0,–15); any point whose *x*-coordinate is zero resides on the *y*-axis.

422. **a.** The *y*-coordinate of every point on the *x*-axis is zero.

423. **b.** Draw a baseball field—its exact shape is irrelevant; only the alignment of the players matter. They form the axis of the coordinate plane. The ball passes the pitcher and veers left of the second baseman; it is in the second quadrant.

424. **b.** Since all of these coordinate pairs have the same y-coordinate they will all sit three units above the x-axis, forming a horizontal line.

425. **a.** Three noncollinear points determine a plane. Choices **b** and **d** are incorrect because •G, •H, and •I do not lie on a common line, nor can they be connected to form a straight line. Caution: Do not assume points are noncollinear because they do not share a common x or y coordinate. To be certain, plot the points on a coordinate plane and try to connect them with one straight line.

426. **c.** First, find the difference between like coordinates: $x_1 - x_2$ and $y_1 - y_2$: $4 - (-2) = 6$. $-5 - 0 = -5$. Square both differences: $6^2 = 36$. $(-5)^2 = 25$. Remember a negative number multiplied by a negative number is a positive number. Add the squared differences together, and take the square root of their sum: $36 + 25 = 61$. $d = \sqrt{61}$. If you chose choice **a**, then your mistake began after you squared -5; the square of a negative number is positive. If you chose choice **b**, then your mistake began when subtracting the x-coordinates; two negatives make a positive. If you chose **d**, then you didn't square your differences; you doubled your differences.

Set 87

427. •A **(1,6).** To locate •A from the origin, count one space right of the origin and six spaces up.

428. •B **(–4,2.5).** To locate •B from the origin, count four spaces left of the origin and two and a half spaces up.

429. •C **(7,0).** To locate •C from the origin, count seven spaces right of the origin and no spaces up or down. This point lies on the x-axis.

430. •D **(0,–3).** To locate •D from the origin, count no spaces left or right, but count 3 spaces down from the origin. This point lies on the y-axis, and x equals zero.

Set 88

For questions 431–434. see the graph below.

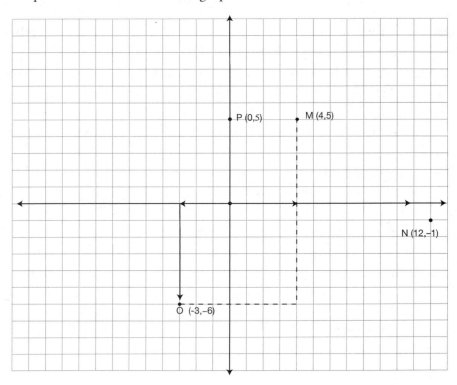

Set 89

435. The horizontal distance between •M and •O is the distance between their x-coordinate: $4 - (-3) = 7$. The vertical distance between •M and •O is the distance between their y-coordinate: $5 - (-6) = 11$.

436. The distance between •M and •O uses the answers from question 435 in the distance formula:

$$d = \sqrt{(x_2 - x_1)^2 + (y_2 - y_1)^2} = \sqrt{(4 - -3)^2 + (5 - -6)^2} = \sqrt{(7)^2 + (11)^2} = \sqrt{170} \approx 13$$

437. *Distance* = **28.** Because these two points form a vertical line, you could just count the number of spaces along the line's length to find the distance between •A and •B. However, using the distance formula: $d = \sqrt{(x_2 - x_1)^2 + (y_2 - y_1)^2} = \sqrt{(0 - 0)^2 + (4 - 32)^2} = \sqrt{0^2 + (-28)^2} = \sqrt{784} = 28$

438. *Distance* = $\sqrt{26}$. $d = \sqrt{(x_2 - x_1)^2 + (y_2 - y_1)^2} = \sqrt{(-1 - 4)^2 + (-2 - (-1))^2} = \sqrt{(-5)^2 + (-1)^2} = \sqrt{25 + 1} = \sqrt{26}$

439. *Distance* = **10.** $d = \sqrt{(x_2 - x_1)^2 + (y_2 - y_1)^2} = \sqrt{(-3 - 7)^2 + (3 - 3)^2} = \sqrt{(-10)^2 + (0)^2} = \sqrt{100} = 10$. Again, because these two points form a horizontal line, you could just count the number of spaces along the line's length to find the distance between •E and •F.

440. *Distance* = **20.** $d = \sqrt{(x_2 - x_1)^2 + (y_2 - y_1)^2} = \sqrt{(17 - (-3)^2 + (0 - 0)^2} = \sqrt{(20)^2 + (0)^2} = \sqrt{400} = 20$. Because these two points also form a horizontal line, you could just count the spaces along the line's length to find the distance between •G and •H.

20

The Slope of a Line

When two points on a coordinate grid are connected, they form a line or line segment which has a slope. The **slope** of a line is the measure of its steepness. Think of slope as the effort to climb a hill. A horizontal surface is zero effort, so horizontal lines have a slope of 0. Steep hills take a lot of effort to climb, so steep lines have greater slopes than gradually tilted lines. Finally, a vertical surface cannot be climbed (without equipment), so vertical lines have an undefined slope (no slope).

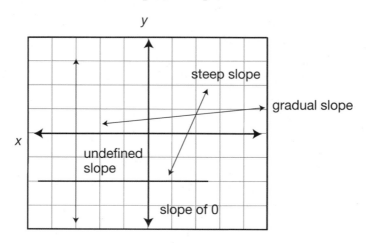

Positive and Negative Slopes

Slope is interpreted similarly to how words are read—from left to right. If a line is angled upward as you look at it from left to right, then it has a positive slope. Conversely, if a line is angled downward when read from left to right, it has a negative slope. In the following coordinate plane, line *l* has a positive slope and line *r* has a negative slope.

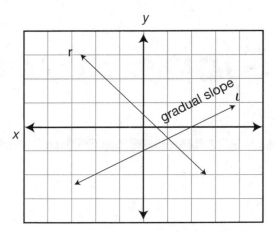

Finding Slope

Slope is represented by a ratio of height to length (the legs of a right triangle), or rise to run. It is written as $\frac{\Delta Y}{\Delta X}$, where ΔY is the change in vertical distance, and ΔX is the change in horizontal distance.

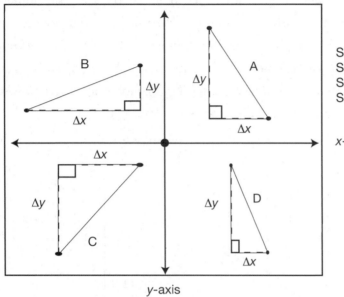

Slope A is negative
Slope B is positive
Slope C is positive
Slope D is negative

The slope between $(-3, 7)$ and $(9, 4)$ can be calculated by plugging them into the following formula:

$$slope = \frac{\Delta Y}{\Delta X} = \frac{(y_2 - y_1)}{(x_2 - x_1)}$$

$$slope = \frac{\Delta Y}{\Delta X} = \frac{(y_2 - y_1)}{(x_2 - x_1)} = \frac{(4 - 7)}{(9 - (-3))} = \frac{-3}{12} = \frac{-1}{4}$$

A slope of $\frac{-1}{4}$ means that for every one unit down, the line move 4 units to the right.

Graphing Slopes

If you are given the coordinate pair for points on a line, and the slope of that line, you can draw the line on a coordinate graph. First, graph the coordinate pair on an axis. Next, starting at that point, plot a second point by following the "map" of the slope: move up or down according the to value of the numerator, and then move left or right according to the value of the denominator. Connect these points to graph the line. If the slope is negative, only *one* of your movements will be negative, *not both*. For example: line *h* has a slope of $\frac{-3}{3}$ and the coordinate pair (5,1) is on line *h*. Starting at (5,1), move three spaces down, and three spaces to the right and plot a second point at (8,−2).

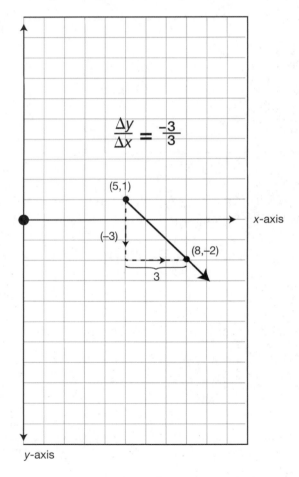

The Slopes of Perpendicular and Parallel Lines

Parallel lines have the **same** slope. Perpendicular lines have **negative reciprocal** slopes. If a slope is $\frac{1}{2}$, a perpendicular slope is -2.

Set 90

Choose the best answer.

441. Pam and Sam are hiking on two separate hills that have the same steepness. If each hill were graphed, they would
  a. have positive slopes
  b. have negative slopes
  c. be parallel lines
  d. be steep lines

442. In American homes, a standard stair rises 7″ for every 9″. The slope of a standard staircase is
  a. $\frac{7}{9}$.
  b. $\frac{2}{7}$.
  c. $\frac{16}{9}$.
  d. $\frac{9}{7}$.

443. Line Q has a slope of $-\frac{1}{9}$. Line T is perpendicular to line Q. Line T will have what kind of slope, compared to line Q?
  a. positive and steep
  b. negative and steep
  c. negative and gradual
  d. positive and gradual

444. Bethany's ramp to her office lobby rises 3 feet for every 36 feet. The incline is
  a. $\frac{36 \text{ feet}}{1 \text{ foot}}$.
  b. $\frac{12 \text{ feet}}{1 \text{ foot}}$.
  c. $\frac{1 \text{ foot}}{12 \text{ feet}}$.
  d. $\frac{36 \text{ feet}}{3 \text{ feet}}$.

445. A plane rises to an elevation of 24,500 feet from 6,300 feet over the course of 30,000 horizontal feet. Which best describes the incline of the plane's course of flight during this elevation gain?

   a. $\frac{91}{150}$

   b. $\frac{24,500}{30,000}$

   c. $\frac{245}{63}$

   d. $\frac{77}{25}$

446. All lines that are parallel to the y-axis have

   a. zero slope.

   b. undefined slope.

   c. positive slope.

   d. negative slope.

Set 91

Find the slope for each of the following diagrams.

447.

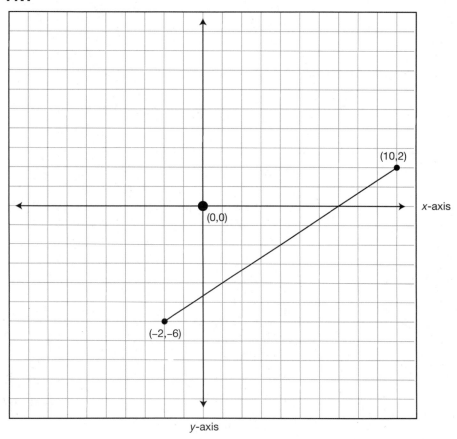

448.

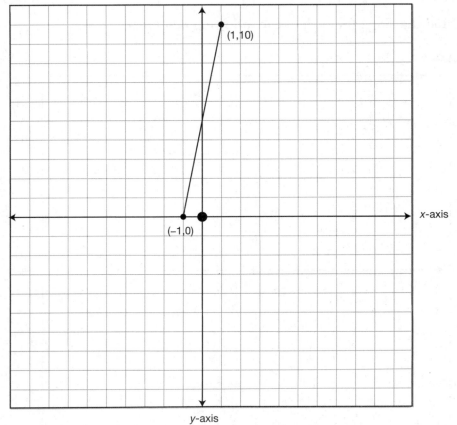

449.

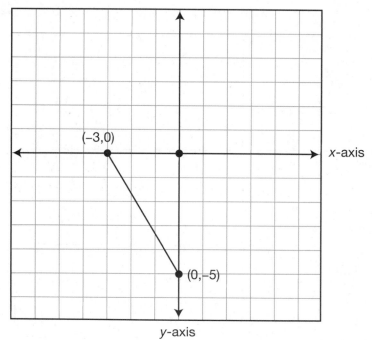

450.

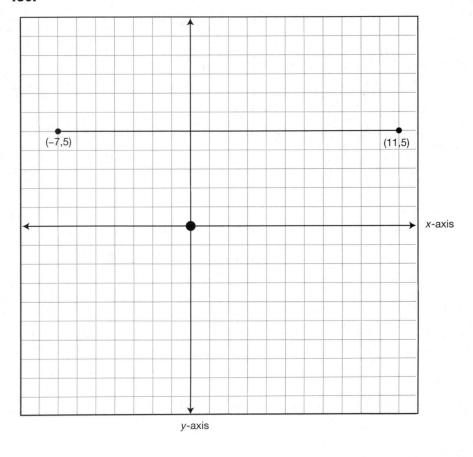

Set 92

Draw each line on one coordinate plane.

451. •M $(0,6)$ lies on line l, which has a $-\frac{5}{2}$ slope. Draw line l.

452. •Q $(-3,-4)$ lies on line m, which has a 3 slope. Draw line m.

453. •S $(9,-2)$ lies on line n, which has a $\frac{1.0}{0.5}$ slope. Draw line n.

Set 93

Use distance and slope formulas to prove the validity of questions 454 through 456.

454. Show that the figure with vertices A (2,−5), B (6,−1), and C (6,−5) is a right triangle.

455. Show that the figure with vertices A (−8,3), B (−6,5), C (4,5), and D (2,3) is a parallelogram.

456. Show that the figure with vertices A (−5,−5), B (−5,−1), C (−1,−1), and D (−1,−7) is a trapezoid.

Answers

Set 90

441. **c.** Because we do not know if Pam and Sam are hiking up or down, answer choices **a** and **b** are not correct. All we know is that the hills have the *same* steepness but we do not know if that is very steep or just slightly steep, so choice **d** is also not appropriate. Since parallel lines always have the same slope, this is the correct choice.

442. **a.** If every step rises 7″ for every 9″, then the relationship of rise over distance is $\frac{7}{9}$.

443. **a.** Since perpendicular lines have slopes that are negative reciprocals of each other, the slope of line T will be 9. Compared to the slope $\frac{-1}{9}$, this slope is positive and steep.

444. **c.** If the ramp rises 3 feet for every 36 feet, then the relationship of rise over distance is $\frac{3 \text{ foot}}{36 \text{ feet}}$. The simplified ratio is $\frac{1 \text{ foot}}{12 \text{ feet}}$.

445. **a.** The plane's elevation gain from 6,300 to 24,500 feet is 18,200 feet. Put this rise over the horizontal "run" of 30,000 feet: $\frac{18,200}{30,000} = \frac{91}{150}$.

446. **b.** All vertical lines have an undefined slope.

Set 91

447. $\frac{2}{3}$. Subtract like coordinates: $-2 - 10 = -12$. $-6 - 2 = -8$. Place the vertical change in distance over the horizontal change in distance: $\frac{-8}{-12}$. Then reduce the top and bottom of the fraction by 4. The final slope is $\frac{2}{3}$.

448. 5. Subtract like coordinates: $-1 - 1 = -2$. $0 - 10 = -10$. Place the vertical change in distance over the horizontal change in distance: $\frac{-10}{-2}$. Then reduce the top and bottom of the fraction by 2. The final slope is 5.

449. $-\frac{5}{3}$. Subtract like coordinates: $-3 - 0 = -3$. $0 - (-5) = 5$. Place the vertical change in distance over the horizontal change in distance: $\frac{5}{-3}$. The slope is $-\frac{5}{3}$.

450. **0 (zero slope).** Horizontal lines have zero slope ($\frac{0}{-18} = 0$).

Set 92

For questions 451 through 453, see the graph below.

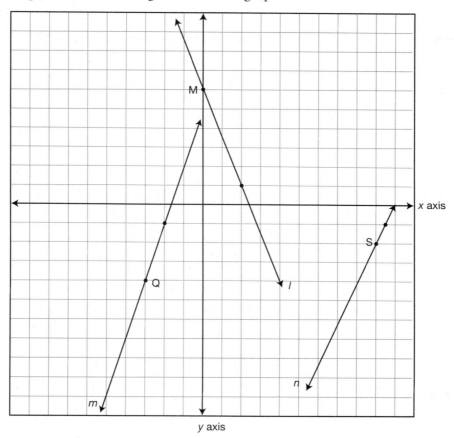

Set 93

454. You could draw the figure, or you could find the slope between each line. The slope of $\overline{AB}$ is $\frac{(-5-(-1))}{(2-6)} = \frac{-4}{-4} = 1$. The slope of $\overline{BC}$ is $\frac{(-1-(-5))}{(6-6)} = \frac{4}{0}$. The slope of $\overline{CA}$ is $\frac{(-5-(-5))}{(6-2)}$, or $\frac{0}{4}$. $\overline{BC}$ is vertical because its slope is undefined; $\overline{CA}$ is horizontal because its slope equals zero. Horizontal and vertical lines meet perpendicularly; therefore ΔABC is a right triangle.

455. Again, you could draw figure ABCD in a coordinate plane and visually confirm that it is a parallelogram, or you could find the slope and distance between each point. The slope of $\overline{AB}$ is $\frac{(3-5)}{(-8-(-6))}$, or $\frac{2}{2}$. The distance between •A and •B is $\sqrt{(-2)^2 + (-2)^2}$, or $2\sqrt{2}$. The slope of $\overline{BC}$ is $\frac{(5-5)}{(-6-4)}$, or $\frac{0}{-10}$. The distance between •B and •C is the difference of the x coordinates, or 10. The slope of $\overline{CD}$ is $\frac{(5-3)}{(4-2)}$, or $\frac{2}{2}$. The distance between •C and •D is $\sqrt{2^2 + 2^2}$, or $2\sqrt{2}$. The slope of line $\overline{DA}$ is $\frac{(3-3)}{(-8-2)}$, or $\frac{0}{-10}$. The distance between •D and •A is the difference of the x-coordinates, or 10. From the calculations above you know that opposite $\overline{AB}$ and $\overline{CD}$ have the same slope and length, which means they are parallel and congruent. Also opposite lines $\overline{BC}$ and $\overline{DA}$ have the same zero slope and lengths; again, they are parallel and congruent; therefore figure ABCD is a parallelogram because opposite sides $\overline{AB}/\overline{CD}$ and $\overline{BC}/\overline{DA}$ are parallel and congruent.

456. You must prove that only one pair of opposite sides in figure ABCD is parallel and noncongruent. Slope AB is $\frac{-4}{0}$; its length is the difference of y coordinates, or 4. Slope $\overline{BC}$ is $\frac{0}{-4}$; its length is the difference of x coordinates, or 4. Slope of $\overline{CD}$ is $\frac{6}{0}$; its length is the difference of y coordinates, or 6. Finally, slope of $\overline{DA}$ is $(\frac{-1}{2})$; its length is $\sqrt{4^2 + (-2^2)}$, or $2\sqrt{5}$. Opposite sides $\overline{AB}$ and $\overline{CD}$ have the same slope but measure different lengths; therefore they are parallel and noncongruent. Figure ABCD is a trapezoid.

21

The Equation
of a Line

Every line on a coordinate plane can be represented with a unique linear equation. This linear equation can be used to generate new points that will be on the line. The same linear equation can also be used to test if given points "satisfy the equation" and sit on the line. Linear equations have two different basic forms, but no linear equation will ever have an x or y variable with any exponent other than 1. (For example, $y = 4x^2$ and $y^3 = 38$ are not linear equations.)

Standard Form of Linear Equations

The first form that linear equations can be written in is called the standard form:

Standard Form of Linear Equations: $Ax + By = C$,
where A, B, and C are all numbers and x and y remain as variables.

To test if a given coordinate pair is on the a line that is in standard form, simply plug the coordinates into the x and y variables in the equation and see if the result is true or false. For example, if you wanted to test to see if the coordinate pairs (5,13) and (4,10) were on the line $-6x + 3y = 9$, plug in each point separately and see:

> (5,13)
> $-6(5) + 3(13) = 9$
> $-30 + 39 = 9$
> $9 = 9$

Since this is a true statement, (5,13) is a solution to $-6x + 3y = 9$, and is on that line.

> (4,10)
> $-6(4) + 3(10) = 9$
> $-24 + 30 = 9$
> $6 \neq 9$

Since this is a false statement, (4,10) is not a solution to $-6x + 3y = 9$, because it is not on that line.

When using an equation to generate points that are on the line, it can be easiest to isolate the x or y variable in the equation. In the following example, the equation $-6x + 3y = 9$ is manipulated to get y on its own so that solutions can easily be found:

$$
\begin{array}{rl}
-6x + 3y &= 9 \\
+6x \qquad &+ 6x \\
\hline
3y &= (9 + 6x) \\
\div 3 \quad &\div 3 \\
\hline
y &= 3 + 2x
\end{array}
$$

The line $y = 3 + 2x$ is the same as the original line, $6x + 3y = 9$, but it is just in a different format. Once y is by itself, it is easy to find points that satisfy this equation and are on the line. This is done by plugging in values for x and then solving for y as is shown in the table below:

$y = 3 + 2x$		
x	y	$3 + 2x$
0	3	$3 + 2(0)$
4	11	$3 + 2(4)$
−1	1	$3 + 2(−1)$

This table shows that the points (0,3), (4,11), and (−1,1) all sit on the line $−6x + 3y = 9$, which is the same as line $y = 3 + 2x$.

Slope-Intercept Form of Linear Equations

The second form for linear equations is the Slope-Intercept form:

Slope-Intercept Form of Linear Equations: $y = mx + b$, where m and b are numbers and x and y remain as variables. This form is particularly useful since m will always be the slope and b will always be the y-intercept. The **y-intercept** is where the line crosses the y-axis. Lines in this form are easy to graph: plot the y-intercept, use the slope to plot two other points, and connect!

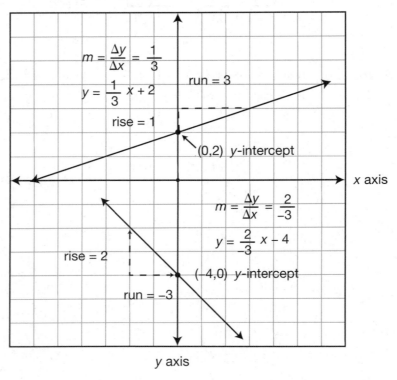

Set 94

Choose the best answer.

457. What distinguishes a linear equation from other types of equations?
   a. its graph is a horizontal line
   b. its graph is a vertical line
   c. its graph is a diagonal line
   d. its variables have exponents of 1

458. The slope and y-intercept of the linear equation $-2x + 3y = -3$ is: (*Hint*: Change this equation into Slope-Intercept form.)
   a. slope $= \frac{2}{3}$, y-intercept $= -1$
   b. slope $= -2$, y-intercept $= 3$
   c. slope $= -\frac{2}{3}$, y-intercept $= -3$
   d. slope $= -2$, y-intercept $= -1$

459. What is the value of b if $(-2,3)$ satisfies the equation $y = \frac{1}{2}x + b$.
   a. -2
   b. -1
   c. 3
   d. 4

460. Which of the following points does *not* lie on the graph for the linear equation $(\frac{3}{2})x - 2y = 14$.
   a. $(1,-\frac{25}{4})$
   b. $(2,-\frac{11}{2})$
   c. $(-2,\frac{17}{2})$
   d. $(0,-7)$

461. Convert the linear equation $4x - 2y = 4$ into a slope-intercept equation.
   a. $y = 2x - 2$
   b. $y = -2x + 2$
   c. $x = \frac{1}{2}y - 2$
   d. $x = -\frac{1}{2}y + 2$

462. •A (–4,0), •B (0,3), and •C (8,9) satisfy which equation?

a. $y = \frac{4}{3}x + 3$

b. $y = \frac{3}{4}x + 0$

c. $y = \frac{3}{4}x + 3$

d. $y = \frac{6}{8}x + 9$

463. Find the missing y value if •A, •B, and •C are collinear: •A (–3,–1), •B (0,y), and •C (3,–9).

a. 1

b. –1

c. –3

d. –5

464. Which line perpendicularly meets line $1x + 2y = 4$ on the y-axis?

a. $y = -\frac{1}{2}x + 2$

b. $y = 2x + 2$

c. $y = -2x - 2$

d. $y = \frac{1}{2}x - 2$

465. Which of the following linear equations is not perpendicular to $\frac{1}{2}x + \frac{1}{4}y = \frac{1}{8}$?

a. $-\frac{1}{4}x + \frac{1}{2}y = -2$

b. $-\frac{3}{2}x + 3y = 21$

c. $2x - 4y = 12$

d. $2x + 4y = 12$

Set 95

466. Which of the following linear equations is parallel to $2x + 3y = 6$?

a. $3x + 2y = 6$

b. $y = 2x + 3$

c. $y = -\frac{2}{3}x - 5$

d. $y = \frac{3}{2}x + 4$

Systems of Linear Equations

A **system of linear equations** is when more than one equation is considered at the same time. When two nonparallel, but coplanar, lines are graphed on the same coordinate graph, they will intersect at exactly one point. This point of intersection is called the **solution** to a system of equations. At this one coordinate pair, there exists an x- and y-value that satisfy *both* linear equations. This (x,y) coordinate pair sits on both lines. **If lines are parallel there will not an intersection or a solution to that system of equations**. In this case, when the system is solved algebraically, the variables will cancel out and you will be left with a false statement, like $4 = -3$. This is how you know that they system is unsolvable because the lines are parallel.

While there are several methods that can be used to find the point that solves a system of linear equations, only two methods will be presented here.

Method 1: Both of the equations are written in $y = mx + b$ form.
In this case, set the two equations equal to one another since they both are equal to "y." Once you do this, you can combine like terms, solve for x, and then plug x back into either equation to solve for y.

> **Example: Find the solution to the system of equations**
> $y = 2x - 20$ and $y = -4x + 40$
>
> Set the equations equal to each other: $2x - 20 = -4x + 40$
> Solve for x: $6x = 60$, so $x = 10$
> Now plug in $x = 10$ to solve for y: $y = 2(10) - 20$, $y = 0$
> The solution is $(10,0)$, which works in both equations.

Method 2: One equation has x or y alone and the other equation does not.
In this case, use substitution to put the equation that has one isolated variable into the other, more complex equation. Then solve for the existing variable and plug that answer back into either equation to solve for the other variable.

> **Example: Find the solution to the system of equations**
> $x = 3y - 12$ and $-2y + 4x = -28$
>
> Sub "$3y - 12$" in for x in the other equation: $-2y + 4(3y - 12) = -28$
> Solve for y: $-2y + 12y - 48 = -28$, so $10y = 20$ and $y = 2$
> Now plug in $y = 2$ to solve for x: $x = 3(2) - 12 = 6 - 12$, $x = -6$
> The solution is $(-6, 2)$, which works in both equations.

Find the coordinate pair that is the solution for each system of equations below.

467. $10x - 9y = 6$
$x = 1 + y$

468. $y = (\frac{2}{3})x + 4$
$y = -1x + 4$

469. $y = (\frac{2}{3})x - 2$
$-4x + 6y = 20$

470. $y = (\frac{3}{4})x - 11$
$y = -(\frac{1}{2})x + 4$

Set 96

Use the line equations below to answer questions 471 through 474. The three linear equations intersect to form △ABC.

$x = 0$
$y = 0$
$y = x - 3.$

471. What are the vertices of △ABC?

472. What is the special name for △ABC?

473. What is the perimeter of △ABC?

474. What is the area of △ABC?

Set 97

Use the line equations below to answer questions 475 through 479. The four linear equations intersect to form quadrilateral ABCD.

$$y = -\frac{1}{3}x - 3$$
$$y = \frac{1}{3}x - 1$$
$$y = -\frac{1}{3}x - 1$$
$$y = \frac{1}{3}x - 3$$

475. What are the vertices of quadrilateral ABCD?

476. Show that quadrilateral ABCD is a parallelogram.

477. Show that diagonals $\overline{AC}$ and $\overline{BD}$ perpendicular.

478. What special parallelogram is quadrilateral ABCD?

479. What is the area of quadrilateral ABCD?

Answers

Set 94

457. **d.** The x and y variables in linear equations can only have exponents of 1. If x or y is squared or cubed, then it is not a linear equation.

458. **a.** Change $-2x + 3y = -3$ into slope-intercept form by adding $2x$ to both sides and dividing everything by 3. The slope-intercept form is $y = \frac{2}{3}(x) - 1$, where the slope $= \frac{2}{3}$ and the y-intercept $= -1$.

459. **d.** Plug the value of x and y from the given coordinate pair into the equation and solve: $3 = \frac{1}{2}(-2) + b$. $3 = (-1) + b$. $4 = b$.

460. **c.** The only point that does not satisfy the linear equation $(\frac{3}{2})x - 2y = 14$ is $(-2, \frac{17}{2})$: $\frac{3}{2}(-2) - 2(\frac{17}{2}) = -3 - 17 = -20$. $-20 \neq 14$.

461. **a.** To convert a standard linear equation into a slope-intercept equation, single out the y variable. Subtract $4x$ from both sides: $-2y = -4x + 4$. Divide both sides by -2: $y = 2x - 2$. Choices **c** and **d** are incorrect because they single out the x variable. Choice **b** is incorrect because after both sides of the equation are divided by -2, the signs were not reversed on the right hand side.

462. **c.** Find the slope between any two of the given points: $\frac{(0-3)}{(-4-0)} = \frac{-3}{-4}$, or $\frac{3}{4}$. •B is the y-intercept. Plug the slope and y value of •B into the formula $y = mx + b$. $y = \frac{3}{4}x + 3$.

463. **d.** The unknown y value is also the intercept value of a line that connects all three points. First, find the slope between •A and •C: $-3 - 3 = -6$. $-1 - (-9) = 8$. $\frac{8}{-6}$ or $\frac{-4}{3}$ represents the slope. From •A, count right three spaces and down four spaces. You are at point $(0,-5)$. From this point, count right three spaces and down four spaces. You are at point $(3,-9)$. Point $(0,-5)$ is on the line connecting •A and •C; -5 is your unknown value for y.

464. **b.** First, convert the standard linear equation into a slope-y intercept equation. Isolate the y variable: $2y = -1x + 4$. Divide both sides by 2: $y = -\frac{1}{2}x + 2$. A line that perpendicularly intercepts this line on the y-axis has a negative reciprocal slope but has the same y intercept value: $y = 2x + 2$.

465. **d.** Change $\frac{1}{2}x + \frac{1}{4}y = \frac{1}{8}$ into slope-intercept form by subtracting $\frac{1}{2}x$ from both sides and multiplying everything by 4 (to get rid of the $\frac{1}{4}$ next to y). The slope-intercept form is $y = -2x + \frac{1}{2}$, which means that the slope is -2. Since perpendicular lines have negative reciprocal slopes, all lines that have a slope of $\frac{1}{2}$ will be perpendicular to $\frac{1}{2}x + \frac{1}{4}y = \frac{1}{8}$. When changed into slope-intercept form, choice **d** becomes $y = -\frac{1}{2}x + 3$ which has a slope of $-\frac{1}{2}$, so this will not be perpendicular to the original equation.

466. **c.** Change $2x + 3y = 6$ into slope-intercept form by subtracting $2x$ from both sides and dividing everything by 3 (to get rid of the 3 next to y). The slope-intercept form is $y = -\frac{2}{3}x + 2$, which means that the slope is $-\frac{2}{3}$. Since parallel lines have equal slopes, all lines that have a slope of $-\frac{2}{3}$.will be parallel to $2x + 3y = 6$. Choice **c**, $y = -\frac{2}{3}x - 5$ is the only equation that has a slope of $-\frac{2}{3}$.

467. **(−3,−4).** Sub "$1 + y$" in for x in the other equation and solve for y:

$10(1 + y) - 9y = 6$
$10 + 10y - 9y = 6$
$1y = 6 - 10$, so $y = -4$
Sub $y = -4$ in for y and solve for x:
$x = 1 + (-4)$, $x = -3$

468. **(0,4)** Set the two equations equal to one another and then solve for x:

$(\frac{2}{3})x + 4 = -1x + 4$
$(\frac{2}{3})x + 1x = 4 - 4$
$(\frac{5}{3})x = 0$
$x = 0$
Sub $x = 0$ in for x and solve for y:
$y = -1(0) + 4$, $y = 4$

469. **No solution: parallel lines.** Sub "$(\frac{2}{3})x - 2$" in for y in the other equation and solve for x:

$y = (\frac{2}{3})x - 2$

$-4x + 6((\frac{2}{3})x - 2) = 20$

$-4x + 4x - 12 = 20$

$-12 \neq 20$. These two lines must be parallel since 12 does not equal 20, which indicates that there is no solution to this system of equations.

470. **(12,−2)** Set the two equations equal to one another and then solve for x:

$(\frac{3}{4})x - 11 = -(\frac{1}{2})x + 4$

$(\frac{3}{4})x + (\frac{1}{2})x = 4 + 11$

$(\frac{5}{4})x = 15$

$x = 12$

Sub $x = 12$ in for x and solve for y:

$y = -(\frac{1}{2})(12) + 4, y = -2$

Set 95

471. **•A (0,0), •B (3,0), and •C (0,-3).** Usually, in pairs, you would solve for each point of interception; however, $x = 0$ (the y-axis) and $y = 0$ (the x-axis) meet at the origin; therefore the origin is the first point of interception. One at a time, plug $x = 0$ and $y = 0$ into the equation $y = x - 3$ to find the two other points of interception: $y = 0 - 3$. $y = -3$; and $0 = x - 3$. $-3 = x$. The vertices of $\triangle ABC$ are A (0,0), B (3,0), and C (0,-3).

472. **$\triangle ABC$ is an isosceles right triangle.** $\overline{AB}$ has zero slope; $\overline{CA}$ has no slope, or undefined slope. They are perpendicular, and they both measure 3 lengths. $\triangle ABC$ is an isosceles right triangle.

473. *Perimeter* **= 6 units + $3\sqrt{2}$ units.** $\overline{AB}$ and $\overline{CA}$ are three units long. Using the Pythagorean theorem or the distance formula, find the length of $\overline{BC}$. $d = \sqrt{3^2 + 3^2}$. $d = \sqrt{18}$. $d = 3\sqrt{2}$. The perimeter of $\triangle ABC$ is the sum of the lengths of its sides: $3 + 3 + 3\sqrt{2} = 6 + 3\sqrt{2}$.

474. *Area* **= 4.5 square units.** The area of $\triangle ABC$ is $\frac{1}{2}$ its height times its length, or $\frac{1}{2}(3 \times 3)$. $a = 4.5$ square units.

Set 96

475. In pairs, find each point of interception:

- A $(-3,-2)$. $-\frac{1}{3}x - 3 = \frac{1}{3}x - 1$. $-\frac{1}{3}x - \frac{1}{3}x = 3 - 1$. $-\frac{2}{3}x = 2$. $x = -3$; $y = -\frac{1}{3}(-3) - 3$. $y = 1 - 3$. $y = -2$.

- B $(0,-1)$. $\frac{1}{3}x - 1 = -\frac{1}{3}x - 1$. $\frac{1}{3}x + \frac{1}{3}x = 1 - 1$. $\frac{2}{3}x = 0$. $x = 0$; $y = \frac{1}{3}(0) - 1$. $y = -1$.

- C $(3,-2)$. $-\frac{1}{3}x - 1 = \frac{1}{3}x - 3$. $-\frac{1}{3}x - \frac{1}{3}x = 1 - 3$. $-\frac{2}{3}x = -2$. $x = 3$; $y = -\frac{1}{3}(3) - 1$. $y = -1 - 1$. $y = -2$.

- D $(0,-3)$. $\frac{1}{3}x - 3 = -\frac{1}{3}x - 3$. $\frac{1}{3}x + \frac{1}{3}x = 3 - 3$. $\frac{2}{3}x = 0$. $x = 0$; $y = \frac{1}{3}(0) - 3$. $y = -3$.

476. In slope-intercept form, the slope is the constant preceding x. You can very quickly determine that $\overline{AB}$ and $\overline{CD}$, and $\overline{BC}$, and $\overline{DA}$ have the same slopes. The length of each line segment is:

$$m\overline{AB} = \sqrt{10}.\ d = \sqrt{(-3 - 0)^2 + (-2 - -1)^2}.\ d = \sqrt{9 + 1}.\ d = \sqrt{10}.$$

$$m\overline{BC} = \sqrt{10}.\ d = \sqrt{(0 - 3)^2 + (-1 - -2)^2}.\ d = \sqrt{9 + 1}.\ d = \sqrt{10}.$$

$$m\overline{CD} = \sqrt{10}.\ d = \sqrt{(3 - 0)^2 + (-2 - -3)^2}.\ d = \sqrt{9 + 1}.\ d = \sqrt{10}.$$

$$m\overline{DA} = \sqrt{10}.\ d = \sqrt{(0 - -3)^2 + (-3 - -2)^2}.\ d = \sqrt{9 + 1}.\ d = \sqrt{10}.$$

477. The slope of a line is the change in y over the change in x. The slope of $\overline{AC}$ is $\frac{-2 - (-2)}{-3 - 3}$, or $\frac{0}{-6}$. The slope of $\overline{BD}$ is $\frac{-1 - (-3)}{0 - 0}$, or $\frac{2}{0}$. Lines with zero slopes and no slopes are perpendicular; therefore diagonals $\overline{AC}$ and $\overline{BD}$ are perpendicular.

478. **Rhombus.** Quadrilateral ABCD is a rhombus because opposite sides are parallel, all four sides are congruent, and diagonals are perpendicular.

479. *Area* **= 6 square units.** The area of a rhombus is its base times its height or half the product of its diagonals. In this case, half the product of its diagonals is the easiest to find because the diagonals are vertical and horizontal lines. $\overline{AC}$ is 6 units long while $\overline{BD}$ is 2 units long: $\frac{1}{2}$(6 units)(2 units) = 6 units.

22

Transformations: Reflections, Translations, and Rotation

What Are Geometric Transformations?

When something is "transformed" it is changed in one way or another. In geometry, points, lines, and even entire polygons can be transformed on a coordinate grid. Sometimes a transformation will result in movement in one or two directions, but the size and appearance of the original figure will remain the same. Other transformations will create a mirror image of an object; other transformations will change a figure's size.

How are Transformations Named?

All transformations use the same method for referring to the original image and naming the transformed image. The original image is called the *preimage* and the new figure is called the *image*. When a point undergoes a transformation, it keeps the same name, but is followed by a prime mark ('). For example, after point Q is transformed, it becomes Q', which is said, "Q prime." Look at the transformation of polygon ABCD in the following figure. Notice that the image it creates is named A'B'C'D'.

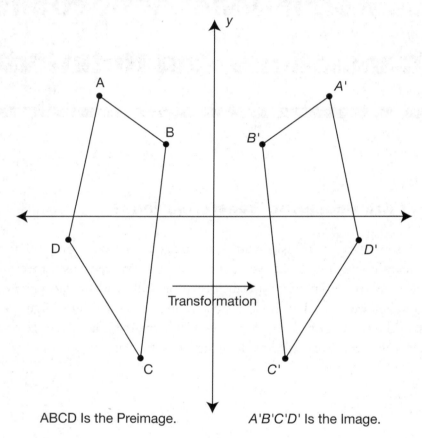

ABCD Is the Preimage.   *A'B'C'D'* Is the Image.

What Is a Reflection?

A **reflection** is a transformation that flips a point or object over a given line. The *shape* of the object will be the same, but it will be a mirror image of itself. It is common that points, lines, or polygons will be reflected over the *x*-axis or *y*-axis, but they can also be reflected over other lines as well such as a diagonal. Polygon ABCD in preceding figure underwent a reflection over the *y*-axis to become A'B'C'D'. In order to reflect a point over the *y*-axis, its *y*-coordinate stays the same, and its *x*-coordinate changes signs. Here are the rules for reflections:

Point A(*x*,*y*) will become A'(–*x*,*y*) after being reflected over the *y*-axis.
Point B(*x*,*y*) will become B'(*x*,–*y*) after being reflected over the *x*-axis.

This figure demonstrates the reflection of point P over the *x*-axis.

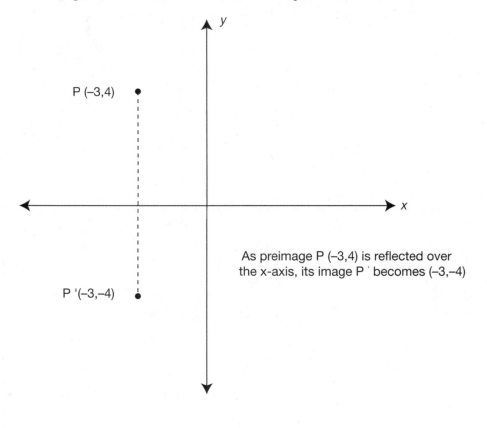

P (–3,4)

P '(–3,–4)

As preimage P (–3,4) is reflected over the x-axis, its image P ' becomes (–3,–4)

What Is a Translation?

A *translation* is a transformation that slides a point or object horizontally, vertically, or in a diagonal shift. When a figure is translated, it will look exactly the same, but will have a different location.

The notation $T_{h,k}$ is used to give translation instructions. The subscript h shows the movement that will be applied to the x-coordinate and the subscript k shows the movement that will be applied to the y-coordinate. These numbers are each added to the x-coordinate and y-coordinate in the preimage.

> **Example:** Translate V(4,–5) using translation $T_{-2,4}$
>
> **Solution:** Beginning with V(4,–5), add –2 to the x-coordinate and add 4 to the y-coordinate: V(4,–5) would be V'(4 + –2), (–5 + 4) = V'(2,–1). This translation is illustrated in the following figure.

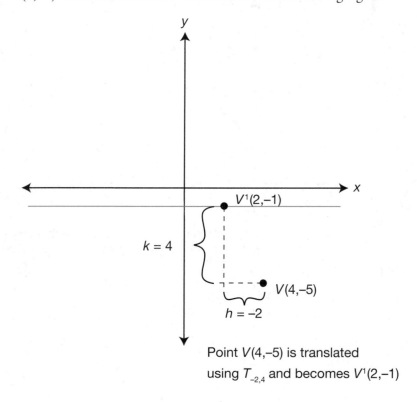

Point V(4,–5) is translated
using $T_{-2,4}$ and becomes V'(2,–1)

Set 98

Choose the best answer.

480. After transforming point W, the new point will be named _____.

481. **True or False:** A transformation will always change the size of an object.

482. **True or False:** A translation will result in an image that is a mirror of the preimage.

483. **True or False:** Reflections only occur over the x-axis or the y-axis.

484. How do you translate a point $R(x, y)$ using $T_{h,k}$?

Set 99

Choose the best answer.

485. After reflecting $N(-3,7)$ over the x-axis, what will the coordinates of N' be?

486. After reflecting $M(6,-1)$ over the y-axis, what will the coordinates of M' be?

487. After translating $L(2,-5)$ using $T_{-3,-4}$, what will the coordinates of L' be?

488. Reflect point $D(5,-2)$ over the x-axis to get point D'. Then reflect D' over the y-axis to get D''. What will the coordinate be of D''?

489. I is reflected over the y-axis to become I'. I' is then reflected over the x-axis to become I''. If I is $(7,2)$ what translation is necessary for I''' to be on the origin?

490. V was reflected over the x-axis to become V'. V' was translated using $T_{-6,2}$ to become V''. If V'' is $(0,-1)$ what are the coordinates of V?

What Is a Rotation?

A rotation is a type of translation that swivels a point or polygon around a fixed point, such as the origin. Rotations are done either clockwise or counterclockwise and are measured in degrees. Most commonly, figures are rotated 90° or 180°. In the following figure ΔKAY in is rotated counterclockwise 90° to form ΔK'A'Y'. You should notice how the following rule was observed in this rotation:

When preimage point A(*x, y*) is *rotated 90° counterclockwise* around the origin, the image point will be A'(-*y, x*).

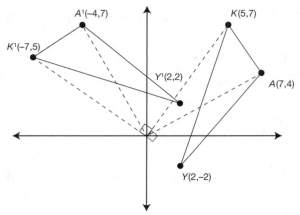

Can you notice what is happening each time a point is rotated 90° counterclockwise?

90° counterclockwise

K(5,7) ⟶ K¹(-7,5)

A(7,4) ⟶ A¹(-4,7)

Y(2,-2) ⟶ Y¹(2,2)

Look at the next figure where points B and G are rotated 90°clockwise. You should notice how the following rule was observed in this rotation:

When preimage point B(*x*, *y*) is *rotated 90° clockwise* around the origin, the image point will be B'(*y*, –*x*).

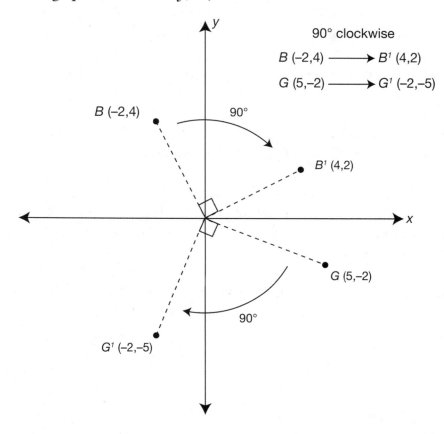

Tranforming Polygons

You have seen a few examples of entire polygons undergoing transformations. To reflect, translate, or rotate an entire polygon, apply the appropriate transformation rule to each individual coordinate pair in the polygon, as was done to ΔKAY earlier in this chapter.

Set 100

Use your knowledge on transformations to answer questions 491 through 495.

491. **True or False:** To rotate a point around the origin, switch the order of x and y and change both of their signs.

492. **True or False:** When U(8,–3) is rotated 90°counterclockwise, the result will be U'(–3,–8).

493. **True or False:** Rotating G(3,3) 90°clockwise will result in the same G' that reflecting G over the x-axis.

494. Preimage points H(–3,–2), P(8,–2), and G(8,5) make a right triangle. The images points of H and G are H'(–2,3) and G'(5,–8). What must P' be in order for ΔH'P'G' to be transformed, but not altered in shape?

495. If ΔDOG is in Quadrant IV, name at least two different combinations of transformations that could move it into Quadrant II.

Set 101

For questions 496 through 501, determine the image point based on the transformation given:

496. Preimage: T(–3,5); Rotate 90° counterclockwise

497. Preimage: S(4,–2); Rotate 90° clockwise

498. Preimage: I(2,7); Reflect over the *x*-axis

499. Preimage: F(–1,–3); Reflect over the *y*-axis

500. Preimage: U(–1,–3); Translate using $T_{-6,5}$

501. Preimage: Z(9,–3); Reflect over the *y*-axis, then rotate 90° counterclockwise to find Z''.

Answers

Set 98

480. After transforming point W, the new point will be named W'.

481. **False.** Only some types of transformations change the size of an object.

482. **False.** Reflections result in mirror images, but translations do not.

483. **False.** Reflections most commonly are over the x-axis or y-axis, but points or figures can be reflected over diagonal lines as well.

484. R' would have the coordinate pair $((x+h), (y+k))$

Set 99

485. N'$(-3,-7)$

486. M'$(-6,-1)$

487. L'$((2 + -3), (-5 + -4)) = $ L'$(-1, -9)$

488. D'$(5, 2)$ and D''$(-5, 2)$

489. $T_{7,2}$. Since I is $(7,2)$, I' would be $(-7,2)$ after a reflection over the y-axis. I'' would be $(-7,-2)$ after $(-7,2)$ is reflected over the x-axis. In order to navigate $(-7,-2)$ to the origin, the translation $T_{7,2}$ is necessary.

490. **(6,3).** Since you are given V'' as $(0,-1)$, reverse the operations of the translation of $T_{-6,2}$ in order to find V': $((0+6), (-1 - 2)) = (6,-3)$ $=$ V'. Since this was the image after a reflection over the x-axis, V must equal $(6,3)$.